美居·奢华
——亚太顶级豪宅设计大赏
BEAUTIFUL HOME LUXURY

聚艺堂文化有限公司 编

中国林业出版社

图书在版编目（CIP）数据

美居·奢华——亚太顶级豪宅设计大赏：汉英对照 / 聚艺堂文化有限公司编
-- 北京：中国林业出版社，2012.5
ISBN 978-7-5038-6533-6

Ⅰ．①美… Ⅱ．①聚… Ⅲ．①住宅－室内装修－建筑设计－中国－图集 Ⅳ．①TU767-64

中国版本图书馆CIP数据核字（2012）第056637号

编写成员：	李应军	鲁晓辰	谭金良	瞿铁奇	朱 武	谭慧敏	邓慧英
	陈 婧	张文媛	陆 露	何海珍	刘 婕	夏 雪	王 娟
	黄 丽	程艳平	高丽娟	汪三红	肖 聪	张雨来	陈书争
	韩培培	付珊珊	高囡囡	杨微微	姚栋良	张 雷	傅春元
	邹艳明	武 斌	陈 阳	张晓萌	魏明悦	佟 月	金 金
	李琳琳	高寒丽	赵乃萍	裴明明	李 跃	金 楠	邵东梅
	李 倩	左文超	李凤英	姜 凡	郝春辉	宋光耀	于晓娜
	许长友	王 然	王竞超	吉广健	马宝东	于志刚	刘 敏

中国林业出版社 · 建筑与家居出版中心

出版：中国林业出版社 （100009 北京西城区德内大街刘海胡同 7 号）
网址：www.cfph.com.cn
E-mail：cfphz@public.bta.net.cn
电话：（010）8322 3051
发行：新华书店
印刷：北京利丰雅高长城印刷有限公司
版次：2012年6月第1版
印次：2012年6月第1次
开本：285mm×285mm，1/12
印张：32
字数：200千字
定价：398.00 元(USD 69.00)

序： 这本集子收录的全是最近几年两岸三地颇有影响力的设计师的作品，作品经过精挑细选，皆为上乘之作。

出这本书的目的只为活跃中国室内设计界的交流，传递一种创新精神；希望能为中国室内设计的发展多注入一点活力。

本书作品的编排皆以作品收录的时间先后为序，编者并未有意调整，所有呈现皆为自然之态，不妥之处望大家批评指正！

CONTENTS/目录

6 **The Luxurious Of Aiax** 爵士白的奢华	106 **Encountering Florence** 邂逅佛罗伦萨
20 **Mission Hills In Residence** 观澜湖高尔夫豪宅	112 **Nobility Residence** 贵族庄园
32 **The Classic Of Beverly Hill** 比华利山的经典	118 **England Feeling** 英伦风情
38 **Repulse Bay Road 56 House** 浅水湾道56号别墅	124 **Gentle Breeze And Leisurely Life** 和风物语
48 **Shenzhen Honey Lake Villa** 深圳香蜜湖别墅	128 **Sumptuous Colors** 流光溢彩
58 **American Dream Villa** 美洲故事别墅	138 **Utopia** 理想国
64 **Noble And Leisureliness** 尊贵雍容	146 **A Singularly Attractive Cavalier** 绝版绅士
80 **The Fashion Of Neoclassicism** 新古典风尚	154 **Duke Territory** 伯侯公馆
86 **Attaching To The Hill** 依恋	166 **Knight Territory** 爵士领地
92 **Love In The Aegean** 情定爱琴海	174 **Adoration** 尊崇姿态
98 **Royal Elegant Residence** 皇家雅居	186 **France Palace** 法兰西宫廷

196	**Precious Collection** 臻品典藏	**306**	**Art Gallery** 艺廊
204	**Sensational Mix & Match** 惊艳混搭	**310**	**Lustrous City** 绚丽之都
218	**Pure Beauty** 纯美乐章	**316**	**Continental Luxury** 欧陆奢华
228	**Sleepless City** 不夜城	**328**	**Shanghai Dragon Lake-Yanlan Hill Villa** 上海龙湖滟澜山别墅
240	**Extreme Noble** 极致尊贵	**336**	**Life Style** 情调
252	**Bright Starry River** 星河璀璨	**344**	**Fragrance Decent** 流香暗袭
260	**French Deep Feeling** 法式浓情	**350**	**Mirroring The Gaudiness** 镜像华美
266	**Sweet Perfume** 馥郁芬芳	**356**	**Archaic Rhymes And New Aristocrat** 古韵新贵
278	**Elegant Modern Luxury** 优雅的现代奢华	**362**	**The Beautiful Of Neutral Classical** 中性古典之美
292	**Mansion Of Refined Scholar** 雅士豪宅	**370**	**The Monumental Beauty** 传世之美
298	**The Beauty Of Floweriness** 绚烂之美		

The Luxurious Of Aiax／爵士白的奢华

项目地点：香港　／项目面积：约450平方米　／设计公司：梁志天设计师有限公司　／设计师：梁志天　／采编：Y·J·Lee

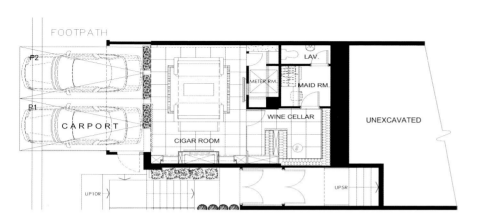

地下室平面布置图

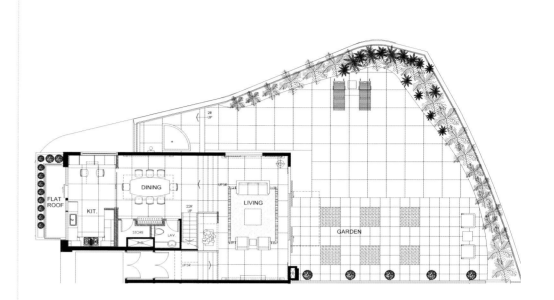

一楼平面布置图

二楼平面布置图

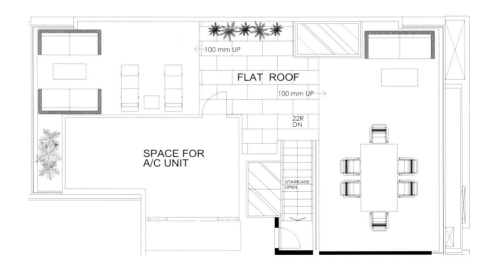

三楼平面布置图

The most characteristic of case is a large area of jazz white stone and extends to each corner as a design clue, which is a rare use in the interior design, it's chiefly because it costs a lot and is hard to control the space and it's hard to obtain the ideal effect if the designer has not deep knowledge and rich practical experience. In this case, the designer uses a large area of the same kind of stone and Intersperse with dark panel on the wall and some simple shapes and some dazzling lamps with metal quality, making the whole space vivid and hierarchical.

When choosing the color of the furniture, the pale colors are better, the decorations are elegant and transparent and the curtains are middle colors, thus the whole space looks more textural. Another clever idea in the case is "borrow view", because the geographical environment is superior and charming, so through the decoration of the windows to make the owner feels like standing before a fantastic picture, it's amazing! In the interior design, the designer doesn't purse the modeling of space too much and the colors are concise and natural. but paying a lot attention to the accessories, that's the trend and fashion of the international design circle.

本案的最大特点就是大面积地使用爵士白石材，并将其作为设计线索延伸到房子的每一个空间，这在住宅室内设计当中极少有人使用，主要是因为第一成本高，第二空间很难把控，要是设计师没有相当深厚的设计功底和丰富的实践经验，一般很难达到理想的效果。在该案当中，设计师在大面积使用同一种石材的同时，再在墙面配以深色的饰面板和一些简单的造型，以及配以一些具有金属质感的壁灯，让整个空间显得生动而又有层次感。
设计师在选择家具的时候尽量以浅色调为主，配饰上要求精致透明，窗帘选择了中间色，这样让整个空间看起来更有质感。本案设计当中设计师还有一个非常高明之处，就是"借景"，因为该项目地理环境优越，室外景观非常宜人，设计师通过对窗户的装饰，让人面对窗户，通过窗外四季景观不同的变化，仿佛看到的是一幅幅精美绝伦的画卷，让人叫绝！在整个空间当中，设计师对空间的造型没有做过多的追求，在色彩上也力求自然简约，但在配饰上却用心良苦，真正做到了多一个嫌多，少一个嫌少的境界。这正是"轻装修、重装饰"国际室内设计界的趋势和潮流。

Mission Hills In Residence / 观澜湖高尔夫豪宅

项目地点：深圳观澜高尔夫别墅区　/项目面积：约500平方米　/设计公司：梁志天设计师有限公司　/设计师：梁志天　/采编：Y·J Lee

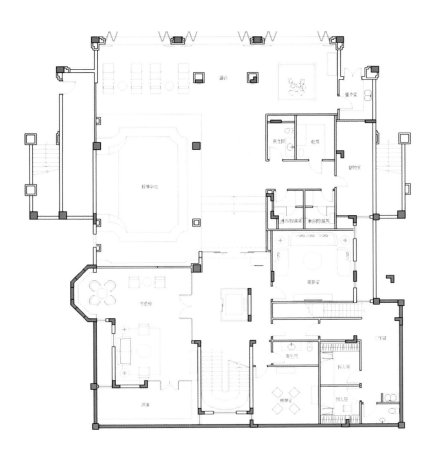

地下层平面图

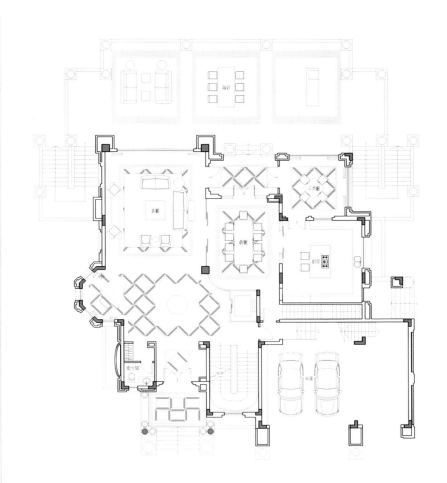

首层平面图

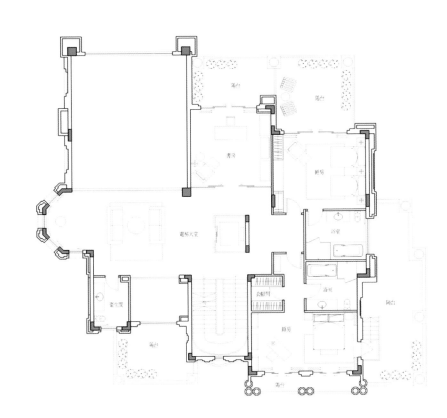

二层平面图

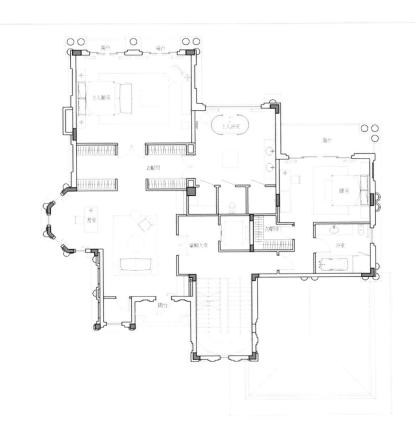

三层平面图

Thinking big goal and win big success. In many villa area, Shenzhen, if we open the window, we can see all the motions of the next-door neighbor. But, if we live in Guanlan lake villa area, when we open the window we can see green grass and blue sky, hear birds singing and smell sweet flowers. We don't need to care about someone will violate our privacy, just enjoy ourselves.

The case uses dainty materials and expertly design, trying to be perfect and flawless, not making the customers have any regrets. The products here are sold to those who expect happy life really. Only those persons with perfectness can readily take a hint when they face such a house.

To Guanlan Lake villa area, the villa is not just space, it is a sort of status symbol, a media of conveying noble smell, a communication way of upper-class society and an attitude of active life.

不将就，才有大成就。深圳的很多别墅，推开窗户就能看到对面人家的一举一动，如果不拉上窗帘，几乎没有隐私可言。在观澜湖别墅群，开窗即景，绿草长天，鸟语花香，自由地舒展身心，尽情地释放自己，不用担心隐私安全。

观澜湖别墅用料考究、设计精美，尽量不留下任何瑕疵，不让业主有任何遗憾。观澜湖的产品是卖给那些真正对生活有所期许的人的，也只有那些"完美主义"的人才能对这样的房子心领神会。

对于观澜湖来说，别墅绝不仅仅只是空间，它是身份荣耀的标签，是传承贵族气息的载体，是上流社会社交的一种方式，也是一种积极向上的生活态度。

The Classic Of Beverly Hill / 比华利山的经典

项目地点：香港新界　　/开发商：恒基兆业地产有限公司　　/设计公司：梁志天设计师公司　　/采编：Y·J Lee

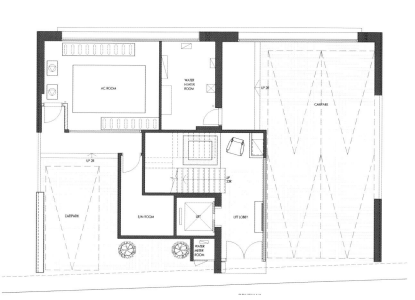

地下室平面设计图

In this case, the designer extends the design with elegance and fashion. The neo-classical is the main design style, which mergers other styles to create a mix match effect. The classical European pattern brings elegant curve to the house, in the golden space, a purple in the sofa makes people feel like stunning at the first sight. In the choice of accessories, the designer pays attention on the dazzling to create the royal manner and adds solemn attitude to the space.
In the Neo-classical style, the simple design of kitchen is fresh and comfortable, realizing the use value at the greatest extent and beautiful and dignified.

在本案中，设计师以典雅与时尚展开设计，总体设计风格以新古典为主，兼并其他风格来营造出一种混搭的效果。经典的欧式花纹为室内带来优雅的曲线，在金色为主色调的空间中沙发的一抹紫色令人一见倾心。在饰品的选择上，设计师注重运用金属感来营造尊贵的皇家气派，为空间更增一份庄重。
在新古典氛围的笼罩中，厨房的简洁设计令人感觉清爽自在，最大程度地实现了厨房的使用价值且美观大方。

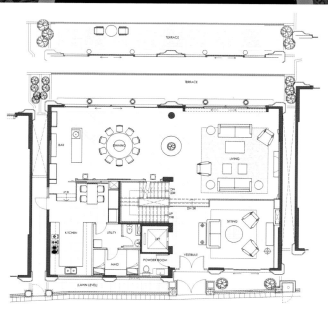

一层平面设计图

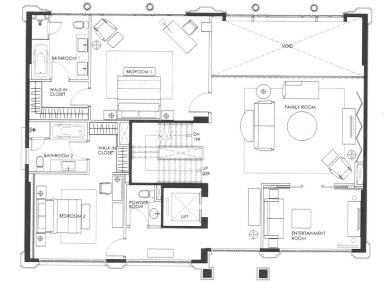

二层平面设计图

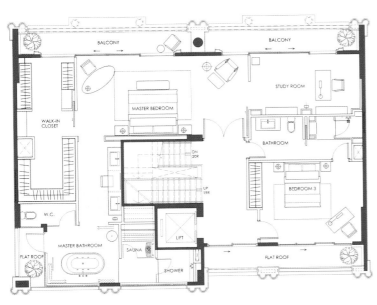
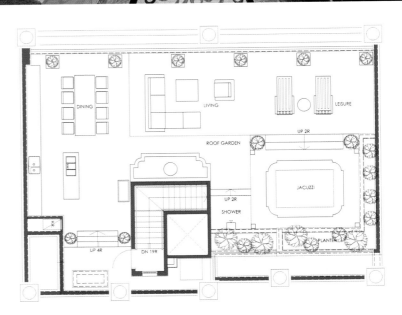

三层平面设计图　　　　　　　　　　　　　　　　　　　露台平面设计图

Repulse Bay Road 56 House／浅水湾道56号别墅

项目地点：香港浅水湾道56号　／项目面积：500平方米　／设计公司：梁志天设计师有限公司　／采编：Y·J Lee

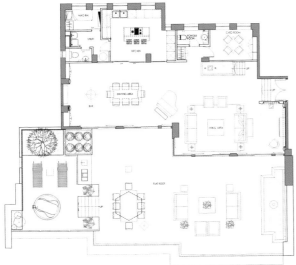

一层平面设计图

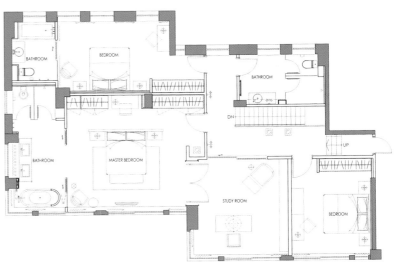

二层平面设计图

Through the designer's mental discernment and elaborated show, the fantastic mansion shows careful attention to details and pursuing tasteful lifestyle. When you get into the gate, you will be shocked by the unique temperament.
The house uses white as the main color and embellishes with grey and silver. The whole house merges the modern classical style set off by vast sea scenery.
The case mixes with modern and ancient design and with crystal droplights showing the content of luxurious demeanor.

对细节的要求和生活品味的追求，透过设计师的独具慧眼和精心演绎，在这个非凡的府邸中尽展新姿。进入大门，便被这座复式房子所渗透出的独特气质深深震慑住。
房子以白色为主调，配以灰色及银色点缀，在辽阔的海景衬托下，整座房子弥漫着一份现代的古典韵味。
本案融合现代与古典设计，加上多盏水晶吊灯及镜面，一派豪宅风范。

Shenzhen Honey Lake Villa／深圳香蜜湖别墅

项目地点：深圳香蜜湖别墅区　／项目面积：800平方米　／设计公司：梁志天设计师有限公司　／采编：Y·J Lee

The interior design of this case is designed by famous Hong Kong designer Steve Leung.
Following the principle of "city centre, honorable mansion", about the theme of "taste and honor" is deepen and detailed . In the interior design the designer pays attention to playing up the tasteful and honorable district atmosphere and the cultural of "relaxation and romantic", which reflects superior and high quality lifestyle.
The designer considers sufficiently the liquidity and Integrity, which embodies the combination of "elegant and quiet "style of architecture, and shows the honorable garden-style accurately, epitomizes freedom and comfort while living in the top–level mansion.

该项目室内设计由香港著名设计师梁志天先生承担。
室内设计遵循"城市中央,尊贵府邸"的项目开发理念,对品味与尊贵这一主题进行有针对性的深化与细化。在进行室内设计时,注重对高品位、尊贵的社区氛围及"休闲、浪漫"文化气息的渲染,体现高品位、高素质的现代豪宅生活方式。
设计师充分考虑流动性、开场性与完整性,并结合体现优雅、平和的"典雅主义"建筑风格及注重内外空间的交流,彰显尊贵的园林风格,集中体现顶级豪宅的休闲与舒适。

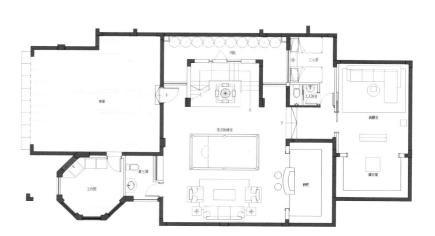

地下层平面图

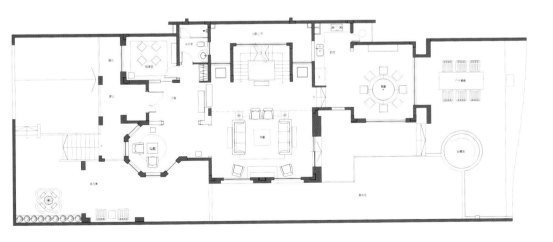

一层平面设计图

二层平面设计图

American Dream Villa / 美洲故事别墅

项目地点：湖南长沙　/项目面积：350平方米　/设计公司：鸿扬家装设计工程有限公司　/设计师：瞿铁奇　/采编：Y·J Lee

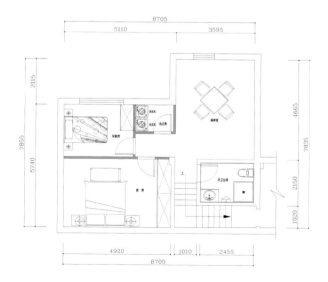

地下层平面图

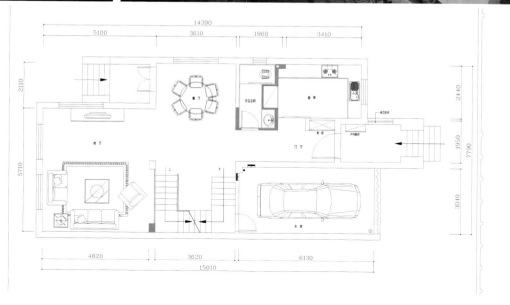

一层平面图

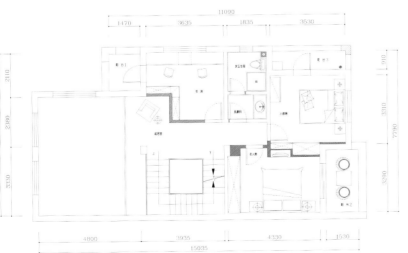

二层平面图

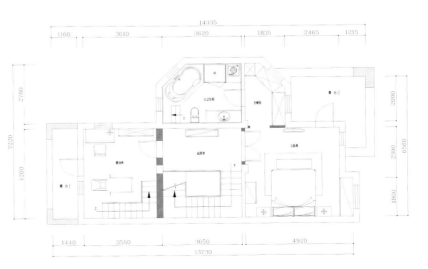

三层平面图

This program is based on the high class villa of american dream besides wanjiali road in changsha,the indoor square is around 350m2 and it's a 3 floor building.The designer finally decided the design plan as modern american style through many time's modification by comnunicating with the owner combined with the villa's own constructing feature.The pointed designing feature in this program is that the disiner knows how to take actions that suit the circumstances,he get more useful function by the lowest cost with the premise of not blighting the impact of space,and simultaneously he do his best on the colour and material,display panel and the choice of wooden stair to make the room be comfortable,harmonious and and unique feature.Mixed with a little white colour formed a room with open-minded , bearing extraodinary,without a slight of unneasiness.The designer used a diversified way of thinking to union the romantic heart longing for the past and the requirement to life of modern people ,compatibled with the luxurious elegance and fashion modernism ,reflecting the aesthetics viewpoint and cultural taste of the post industrial time.

该项目位于长沙市万家丽路旁边的美洲故事高档别墅区内，室内面积约为350平方米，共三层。设计师通过和业主的沟通，结合别墅本身的建筑特点，将设计的方案定位为现代美式风格，后几经修改，终于得以成形。该项目最大的设计特点就是设计师懂得因地制宜，用最小的成本在不影响空间效果的前提下获得更多的实用功能，同时设计师在空间色彩和材质上也下足了功夫，大块面使用饰面板以及选用木质楼梯，力求让空间显得温馨、和谐而又有独特的气质，少量白色糅合，使色彩看起来明亮、大方，使整个空间给人以开放、宽容的非凡气度，让人丝毫不感局促。设计师用一种多元化的思考方式，将怀古的浪漫情怀与现代人对生活的需求相结合，兼容华贵典雅与时尚现代，反映出后工业时代个性化的美学观点和文化品位。

Noble And Leisureliness／尊贵雍容

项目地点：广州市天河区珠江新城马场路28号　／项目面积：约263平方米　／开发商：广州富力地产股份有限公司　／采编：Y·J Lee　鲁晓辰

The design inspiration of this case comes from the modern Neoclassicism of Italy. The whole flat takes the old classic bosom as keynote but doesn't lose its modern feeling. Neoclassicism furniture and the modern shape space reflect mutually and the whole space is full of high taste.

The light indoor is soft and the unique style of decoration creates high-tasted comfortable atmosphere. The furniture made of wood works in concert with the carpet painted with flowers. Life vegetation all over the room and the natural landscape outdoor make people feel like living in the nature again. It shows the high taste and steady of Neoclassicism in the details and makes people feel extravagant, fashionable, mature and more tenderly.

本案的设计灵感来源于意大利的现代新古典主义，整套样板房以经典怀旧为基调，但不失现代感。新古典风格家具与空间的现代造型相互辉映，使整个空间渗透着浑然不觉的高雅气质。

整个室内光线柔和，造型精巧别致的小饰品在其笼罩下并无喧宾夺主之感。木质家具与花朵图案的地毯相互呼应，绿色植物生机傲然地穿插各处，室内室外自然景观给予客户回归深林的动人情怀。一处处不经意透露的细节更升华新古典主义的高贵、沉稳，令人感觉奢华却不失时尚，成熟而倍感温馨。

平面设计图

The Fashion Of Neoclassicism／新古典风尚

项目地点：广东佛山　／项目面积:440平方米　／设计公司：广州市韦格斯杨设计有限公　／采编：Y·J Lee　鲁晓辰

The case is a three-story villa. It takes black, white and grey as the hues inside, showing the neoclassical style heartily. Outstanding painting and refined soft furniture form delightful contrast. Decoration with stainless steels also provides astounding feelings.

Selected high ceiling designs widen the space. French windows not only extend the sense of visional space, but also make for lighting. Marble wall keep consistence with the theme color. Meanwhile, oversized gorgeous chandelier twinkles throughout the whole living room. On the contrast, study room is elegant decorated. Tailor-made bookcase is very useful. Without complicated design, it shows nobiliary and steady.

本案为三层叠加复式别墅，室内以黑白灰三色为主调尽情展现新古典主义风格。经典的西式传统造型轮廓中，突出画块之间的造型，精炼的线条感与软式家具相映成趣，饰品点缀不锈钢材料，令客户有耳目一新的感受。

挑高设计使空间变得开阔，而大幅玻璃开窗不仅扩大了视觉空间，更便于采光。大理石背景墙色彩与主色调一致，超大尺寸华丽吊灯将光线发散至各处，整个客厅顿时熠熠生辉。

书房布置精炼优雅，线条感十足。特别定制的书橱以实用为主，摒弃了繁复夸张的设计，显露出沉稳姿态。

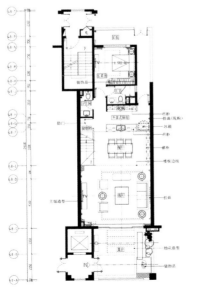

一层平面设计图

二层平面设计图

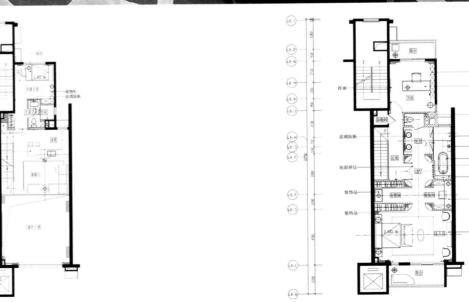

三层平面设计图

Attaching To The Hill／依恋

项目地点：广州北路花都九龙湖国际社区 项目面积：536平方米 /开发商：新鸿基地产发展有限公司 /采编：Y·J Lee 鲁晓辰

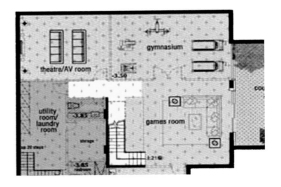

一层平面设计图　　　　　　　二层平面设计图　　　　　　　三层平面设计图　　　　　　　四层平面设计图

The Jiulong Lake international community where the Jiulong Lake sits is a national AA level large landscape ecological resort. It is embraced by more than 30000 acres of forest park, and thus gets its name. The Yueyuan Group-Purchase provides more than 240 sets of contemporary and contracted style villa, which is mainly composed of single-family villas, and double-famliy villa and low density house products. All the villas are north-and-south-oriented designed, lighting and comfort relying on natural advantages.

The Phase A of the Yueyuan Group-Purchase is adopted the design of wide face and big width room, using many French window glass screens, to assure that the customs can appreciate the beautiful scenery of mountains and rivers indoor, at the promise of sufficient lighting.

The interior design of H House Style "Yiluan" introduces the gold and purple into the main hue, fitting with graceful and beautiful furniture and bright crystal glass decoration, to create a unique boundless temperament. Elegant adornment is placed indoor everywhere, with the close combination of open-outdoor-and-private-indoor style, providing clients with unique perfect experience.

The designer highly extracts the European type elements, using a expression of new classicalism, to built a brand new European mansion of high visual impact for the habitants.

玖珑湖所在的九龙湖国际社区，为国家AA级大型山水生态度假区，周围更被3万亩的森林公园所环抱，依恋之名由此得来。悦源组团共提供240余套现代简约风格的别墅，以独栋别墅为主，配以双拼别墅及低密度洋房产品。所有别墅均为南北向设计，依仗自然优势来保证采光和舒适。

一期悦源组团采用的是大面宽、大开间设计，较多地运用了全落地玻璃幕围，在保证室内通透的前提下，山水如不请自来般呈现于眼前。

H户型"依恋"室内设计以金色和紫色为主色调，搭配雍容华美的家具及璀璨耀眼的水晶玻璃饰品，营造出不同凡响的大家气质。雅致的装饰摆件点缀于室内各处，室外的开放与室内的私密紧密结合，为客户提供别具一格的完美体验。

设计师通过对欧式元素的高度提炼，并采用新古典主义的表达方式，为居住者营造了一个耳目一新的、极具视觉冲击力的欧式大宅。

Love In The Aegean／情定爱琴海

项目地点：广州从化　／项目面积：650平方米　／设计公司：广州市韦格斯杨设计有限公司　／开发商：广州珠光房地产开发有限公司　／采编：Y·J Lee　鲁晓辰

This case is a three-layer single-family villa. It has private front and back garden with a hollow courtyard in the middle, thus lighting and ventilation effect are extremely good.

Grand and luxury adornment creed style is that of the design. Indoor use lines and geometry, fully displays the space extensibility .The grid form floor of lounge mutually reflects the design of ceiling, and the round gold foil top of the dining room corresponds with the floor style. Classic symmetric feature shows in the adornment. The exquisite geometrical design used in the large carpets and curtains is based on square, diamond and triangle, let whole space full of gradation and metrical sense.

The indoor main hue is gold, with the collocation of luxury Europe type furniture, the whole room sends out a dense luxuriant temperament. Brunet sofa decorates with gold lace, nostalgic oil painting with pure gold frame, classical black golden-edge tea table in the sitting room, the whole design echoes with the details, elegant noble flavor is coming against your face .

本案为三层独栋别墅，前后私家花园，中间配有中空庭院，采光及通风效果极好。

隆重而奢华的装饰主义风格是本案的设计风格。室内运用线条和几何图形，充分展示空间的延伸性。会客厅格子形的地板与天花板的设计相互辉映，餐厅的原型金箔顶部与地板浓烈对应，装饰主义风格中的经典对称展现无遗。大块地毯及窗帘所采用的精致的几何图案，以方形、菱形和三角形为基础，让整个空间充满层次感及韵律感。

室内的主色调为金色，搭配豪华的欧式家具，整个房间散发出华丽而浓重的宫廷气质。深色沙发缀以金色花边，复古油画配上纯金色画框，客厅古典黑色描金边茶几，整体与细节的契合、呼应耐人寻味，优雅的贵族气息扑面而来。

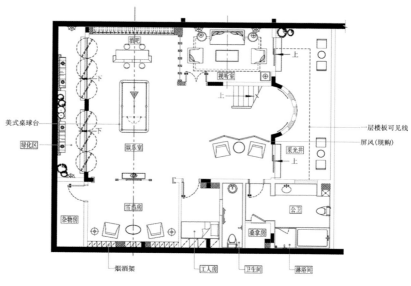

地下层平面设计图

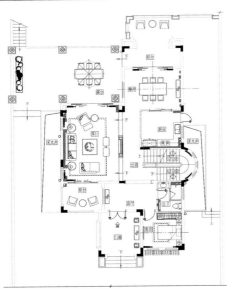

一层平面设计图

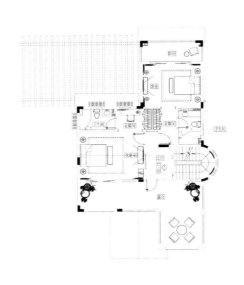

二层平面设计图

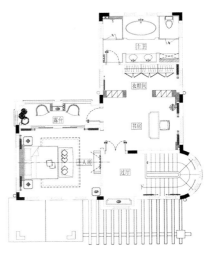

三层平面设计图

Royal Elegant Residence／皇室雅居

项目地点：广州从化　／项目面积:650平方米　／设计公司:广州市韦格斯杨设计有限公司　／开发商：广州珠光房地产开发有限公司　／采编：Y·J Lee　鲁晓辰

This case is developed by Pearl River Real Estate, located by the Liuxi River in Conghua, featured by the senior single-family villas of four-floor and a pure beautiful and gentle Aegean Sea theme interior design, to create a fantastic living space for customers.

Bright color of the blue sea and sky sublimates the human's most pure feeling. White wall face, horseshoe-style doors and windows, refined arch door, soft line and classic blue maize ornament, air casual slowly circulation, the romantic breath of the age before the Renaissance flows.

Interior decoration adopts natural white as the main hue, dedicatedly forged iron works keep the elegant and practical feature of Mediterranean style. In accordance with blue curtain and yellow light, it makes a person as if he is in the white village of the northern coast. Sofa is simple and practical, whose round pattern is sending out the comfortable feeling. The colorful lights abandon the glaring nature, keep the nature fresh and pure. This case design makes full use of space, it does not show constraint or break the atmosphere, and liberates the open free space, letting a person always feel the Mediterranean style furniture sending out the old a rural breath and culture grade.

本案为珠江地产开发的位于从化流溪河畔的高级独栋四层别墅，室内设计以纯美与柔和的爱琴海为主题，为客户营造一个如梦似幻的居住空间。

碧海与蓝天明亮的色彩，升华了人类最纯粹的情愫。纯白的墙面、马蹄状的门窗、雅致的拱门、柔和的线条以及经典的蓝黄色点缀，空气在不经意间缓缓流转，室内顿时洋溢着文艺复兴前的浪漫气息。

室内装饰以自然的白色为主基调，精致锻打的铁艺饰品保持了地中海风格的优雅与实用性，配合蓝色的窗帘及黄色灯光，使人仿佛置身于蔚蓝海岸旁的白色村庄。沙发简约实用，浑圆的造型散发着舒适的芬芳。多姿的灯光摒弃了炫目的姿态，还原了大自然的纯净清新。本案设计充分利用空间，不显局促、不失大气，解放了开放式自由空间，让人时时感受到地中海风格家具散发出的古老尊贵的田园气息和文化品位。

Encountering Florence／邂逅弗洛伦萨

项目地点：杭州西湖九溪风景区　／项目面积:400平方米　／设计公司：上海全筑建筑装饰设计公司　／开发商：绿城集团　／采编：Y·J Lee　鲁晓辰

Green City-Jiuxi Rose Garden is located in Hangzhou Zhijiang National Tourist Holiday Resort, staying by Qiantang River, Wuyun Moutian, Jiuxi and Yunxi, close to Six Harmonies Pagoda, the natural environment is incomparable. This case is a set of private villa , Mediterranean style is its theme of design .

Classical ceramic tile on the wall, round and smooth ironware,and exquisite mosaic surfacing archway, thus setting up a delightful fresh felling of Mediterranean littoral. And the pure color sculpture of wall filled with European taste, making people feel like standing over the sculpture in Florence, facing the Revival of Letters. The combination of pure color makes Mediterranean be an unique style, living room, saloon and bedroom of this case embellish respectively with purple and red, then make people astonished when they see it, meanwhile bringing the weald feeling of the earth . The den adopts the wood quality decoration, letting everything come to a quality suggestive of poetry or painting of nature.

The designer integrates the feeling of the Renaissance into this case, trying hardly to give a newly living experience, opening up a full of warm and soft warm country house and free lifestyle to their customs.

绿城-九溪玫瑰园位于杭州之江国家旅游度假区内，在钱塘江、五云山、九溪和云栖之间，邻近六和塔，自然环境无与伦比。本案为一套独栋别墅户型，地中海风格是其设计主题。
墙面古朴的瓷砖、浑圆的铁艺饰品以及精致的马赛克拱门，搭建起地中海沿岸的清新印象。墙面欧洲风味的纯色浮雕艺术造型感十足，让人仿佛流连在弗洛伦萨的雕塑前，与文艺复兴遥相对望。纯美色彩的组合令地中海风格别具一格，本案的客厅、会客厅及卧室分别点缀以紫色、红色，让人惊鸿一瞥中目眩神迷，而后又感受到大地般的旷野感。书房采用的木质装潢，让一切均回归自然的诗情画意。
设计师将文艺复兴时期的艺术感融入本案，力求给客户一种全新的居住体验，将充满温暖柔和的乡村居所与随性的休闲生活方式尽情展现。

Nobility Residence／贵族庄园

项目地点：福州晋安区　／项目面积：约350平方米　／设计公司：福建国广一叶建筑装饰设计工程有限公司　／开发商：广州富力地产股份有限公司　／采编：鲁晓辰

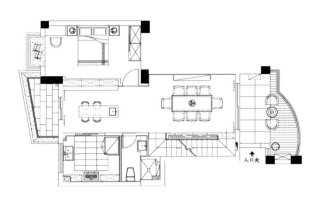

一层平面设计图

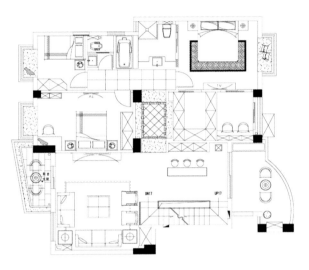

二层平面设计图

三层平面设计图

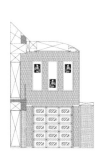

立面图一

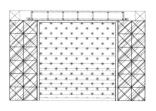
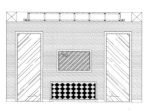

立面图二

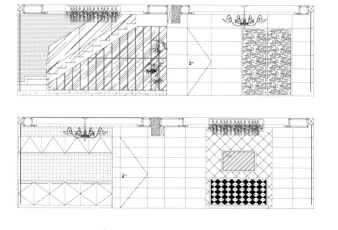

立面图三

This case is a three-story villa example room. The designer uses the modernistic style, blending in European luxury feeling, and strives to provide clients with unique noble enjoyment.

In the sitting room the space is extended high, coordinated with three-dimensional large area bead shade crystal droplight and Europe type sofa, fully releasing palace magnificent atmosphere. At the same time, taking the mysterious blue and abstract square as the ornament, it not only builds composed temperament with a grey-black-and-white color, but also gives the customers luxurious and elegant feelings, as if they can own a exclusive comfortable space in the prosperous city.

Luxury interior decoration doesn't mean impractical. It feels intense space combination, with smooth indoor transition, and the shape of the roof matches the light color. Space is not confined to the hard wall, but pays more attention to the mutual relationship of the function of visiting, dining, learning and sleeping. The open kitchen sits by the stairs bravely uses pure white walls, and dining room not far away uses all black top and black-white desks and chairs. Extremely personality and tension shows in this combination, with so much space flexibility and compatibility.

本案是一套别墅样板房，共三层。设计师采用现代主义风格，融入欧式奢华感觉，力求为客户带来别具一格的高贵享受。

客厅处将空间挑高设计，搭配立体大面积珠帘水晶吊灯和欧式沙发，尽情释放宫廷华丽氛围。同时，以神秘的蓝色和抽象的方形点缀，不仅勾起黑白灰三色搭建的沉稳气质，更令人感觉到整个空间的奢华、高雅，在繁华的都市里仿佛拥有了一个独享的自在空间。

室内装饰虽雍荣华丽，但实用性毫不逊色。空间组合感强烈，不同用途的室内流畅过渡，从屋顶形状、到灯光颜色互相渗透，空间划分不再局限于硬质墙体，而是更注重会客、餐饮、学习、睡眠等功能的逻辑关系。楼梯旁的开放式厨房内大胆采用纯白墙壁，不远处的餐厅则采用全黑顶部及摆放黑白桌椅，整个搭配极具个性与张力，空间表现灵活且兼容。

England Feeling / 英伦风情

项目地点：上海　/项目面积：300平方米　/设计公司：上海全品室内装饰配套工程有限公司　/开发商：上海安东房地产发展有限公司　/采编：Y·J Lee　鲁晓辰

This case is a set of superposition villa, the designer uses British style design, introducing the elegant tastes of Great Britain into modern life.
The British style is adhering to the classical symmetry and harmonious principle, paying attention to detail processing, and trying to build a gorgeous image of delicate temperament. In this case, through the symmetrical beauty the French window and the Europe type sofa reveals an elegant atmosphere, and the golden arc curtain passes the whole warm feeling.
Yellow as the main indoor hue, in each room tables, chairs and corridor stairs are all wood-made. The whole hue is unified. The carpet with small flowers accords to the British classic appeal, the complexity of adornment is abandoned and is taken place by green plants, oil paintings and bringing us living and interest, and creates colorful and practical English aristocrat life experience.

本案户型为一套叠加别墅，设计师采用英式设计，将大不列颠的优雅情趣融入现代生活。
英式风格秉承古典主义对称与和谐的原则，注重细节的处理，试图营造一种华丽精致的形象气质。本案中，通过对称美，以对角线为中轴的落地窗和欧式沙发彰显出扑面而来的大气优雅，而金黄色的弧形窗帘则传递着整体的温馨感受。
室内的主基调为黄色，而各室桌椅、过道楼梯均采用木质，整体色调统一。碎花的地毯图案秉承着英式的古典韵味，室内摒弃了纷繁复杂的装饰品，由绿色植物、油画给房间带来生机趣味，创造纷呈而实用的英国贵族生活体验。

Gentle Breeze And Leisurely Life／和风物语

项目地点：上海市青浦区　／项目面积：约400平方米　／开发商：上海万科房地产有限公司　／采编：Y·J Lee　鲁晓辰

This case is an example room of the villa, It takes British rural life as the blueprint and depicts rural villa life in central England.

The essence of Vanke Jingyuan comes from the classic "heart of the England"--the "Cotswolds" style architecture art, with expensive building elevation, giving the building a low-key, sophisticated temperament, revealing the beauty of elegance, and restoring pure rural scenery of England.

Indoor design adopts a large number of wood furniture with warm color lamplight, best matching the delicate carpets, and making space present a rich and unified color level. Luxury English aristocratic dresses and all kinds of decorations build a classic visual effect, making everything look so beautiful and moving. Children room uses the popular navy favor, blue sheet with white flower echoes with wall color line, showing comfortable and fashionable. The whole set of example room enjoys a unified style, and perfectly builds a dream-like English gentleman temperament.

本案为别墅样板房，以英式田园生活为蓝本，描摹出英格兰中部的田园别墅生活。

万科晶源源自英格兰之心的经典——"科茨沃尔德"式建筑艺术，以厚重不菲的建筑立面，赋予建筑低调、沉稳的气质，彰显俊逸之美，还原纯正英伦田园的风景。

室内大量地采用木质家具，配合暖色灯光，与精致地毯相得益彰，空间呈现出丰富而统一的色彩层次。华丽的英式贵族服饰与各色摆设营造出古典的视觉效果，使一切看起来都那么唯美动人。儿童房采用了热门的海军风，蓝底白花床单与墙壁彩色线条呼应，时尚中充满舒适感。整套样板房风格统一，完美地搭配出世人梦中的英国绅士气质。

Sumptuous Colors / 流光溢彩

项目地点：温州　／项目面积:356平方米　／设计公司：温州麦高室内设计事务所　／采编：鲁晓辰

The highlight of the design in this case is how the designer cleverly mixes the classic beauty and modern elements together.
Nostalgic brunet sofa expresses the veil of the classical atmosphere, bright baroque crystal light sends out charming spirit of opera, and the diamond mirror on the TV set wall shines to extend a higher space, all of the above involves the customers into the dreamy palace Waltz. But with the creative use of fashionable silver foil, animal-line pillow and all kinds of simple line furniture, modern breath sneaks into the indoor space, releasing a dense luxuriant feeling, the whole indoor space is filled with light rhythm and elegant.
Compared to the colorful sitting room, the bedroom is more emphasis on comfort and peace. Plain colour archaize stripe wallpaper combines with concise-lines furniture making people forget everything prosperous blundering, and returning low-key brilliance.

本案的设计亮点在于设计师将古典美与现代元素巧妙融合。
复古深色的沙发大气地挑起古典的面纱，璀璨的巴洛克水晶灯散发歌剧般的魅惑，电视背景墙的菱形镜面映照挑高空间，一切将时空卷入如梦似幻的宫廷圆舞。但是，随着时尚的银箔，动物纹的靠枕以及各类线条简单的家具的大胆使用，现代气息见缝插针地渗入室内，释放了浓烈的华丽感受，整个室内节奏轻盈却不失优雅。
相比起客厅的流光溢彩，卧室更多的则是强调舒适宁静。素色仿古条纹壁纸搭配简洁线条的家具，使人忘却一切繁华浮躁，回归低调柔和。

Utopia / 理想国

项目地点：西安市　/项目面积：230平方米　/设计公司：PINKI(品伊)创意机构刘卫军设计师事务所　/开发商：西安深鸿基房地产开发有限公司　/采编：鲁晓辰

European culture contains rich artistic heritage, as brilliant as stars. And its innovative design thought and exalted essence is long time favored by artists. This case interior design is filled with pure and smooth artistic temperament, making the inhabitants feel like lingering among the clouds fairyland, far away from the dusty world.
In this case the sitting room give priority to white combined with black simple lines. Elegant and clear art favor is coming against you face. The designer reconstructs the original garden veranda into restaurants, letting the inhabitants enjoy the special experience of having a dinner at the perspective of 270 degree . Black-white decorative pattern wallpaper, concise cloth art sofa, pure white noble triangle piano, glittering and translucent crystal droplight and quietly blossoming elegant flowers, and so on, both the overall and the details are giving the inhabitants quietly elegant and noble experience. The feelings of the owner pursuing the beauty of art combined with contemporary style provide us with strong feelings of the perfect collisions of deep cultural background and personality.

欧洲文化蕴含着丰富的艺术底蕴，如繁星般浩渺无垠，其开放、创新的设计思想及尊贵的姿容一直受到艺术家的青睐。本案室内设计充满流畅纯净的艺术气质，居住其中感觉如云端仙境，一切世俗尘埃俯首称臣。
本案客厅以纯白为主基调，配合黑色简单线条，高雅、明快的艺术气息如清风拂面。设计师将原有的花园凉台改造为餐厅，270度开阔视角下的用餐感受必定非同寻常。黑白花纹图案的墙纸、简洁的布艺沙发、纯白高贵的三角钢琴、晶莹剔透的水晶吊灯及淡雅盛放的花朵等等，整体到细节均带来清醒淡雅却又高贵精致的体验，主人追求至美艺术的情怀与现代感结合，使人强烈感受浓厚的文化底蕴与个性的完美碰撞。

平面布置图

A Singularly Attractive Cavalier/绝版绅士

项目地点：西安市　/项目面积：220平方米　/设计公司：PINKI（品伊）创意机构刘卫军设计师事务所　/开发商：西安深鸿基房地产开发有限公司　/采编：鲁晓辰

In this case the indoor structure is relatively normal, giving us a sedate feeling. It takes red, yellow and blue as the main hue to show the owner's affection towards life and his special insight.

The preference towards classical style favor makes the designer choose nostalgic drawings decorated with square frames, primitive nostalgic Mosaic to compose an elegant atmosphere. The sitting room is permeated with brunet wood feeling, the top droplight which is not costly but full of a design feeling softly shining everything, dark red wallpaper echoes the red soft skin bedside at a distance, both the sitting room and the bedroom inherit the artistic gene from the history. Slamming the glaring luxury, it perfectly combines the cold and warm hue, contracted and classic features depend on each other…All these seem to be a dismissing classical ball, breathing and inspiring the moving exotic ambience.

本案的室内结构较为方正，整体感觉周正稳重，主要以红、黄、蓝三色为主色调，彰显主人对生活的火热情怀和非凡见地。
对经典古典风格的钟情，使设计师选择方形画框装点下的复古图纸、古朴怀旧的马赛克，拼凑出溢于言表的优雅气息。客厅洋溢着深色的木质感，毫不奢华却设计感十足的顶部吊灯柔和注视着一切，暗红色壁纸与红色皮质软包床头遥相呼应，客厅及卧室不经意流露着艺术感的一脉相承。摒弃了金光炫目和纸醉金迷，房间中冷暖色调相互交错，简约与古典彼此依存，一切仿佛是散场的古典舞会，呼吸吐纳着动人的异国气息。

平面设计图

Duke Territory／伯侯公馆

项目地点：广州市番禺区　/项目面积：约380平方米　/开发商：广州番禺雅居乐房地产开发有限公司　/采编：Y·J Lee

The design of this case adopts the Northern Europe style, in which local cold and wet oceanic climate makes the designer decorate the room with the warm yellow hue indoor purposely.

The indoor design obviously shows the preference of designer to use plane figures, especially square and circular. The sitting room floor used small squares and the simple outline sofa with squared sculpt bring to us with home feeling. In addition, the ceiling of the sitting room echoes with the circular crystal droplight in the dining-room, shedding light on the primitive warm-colored table and surrounded-style seats, concomitant with the round shape modern arts, and make the whole room resound the inspiration of art.

In the quiet and beautiful space, whether purplish red carpet and bright-coloured flowers, or a pair of orange poster, all beats surprising modern elements, which has become a highlight of this case. The French windows in each room not only considerably keep up abundant natural lighting but also make indoor home and outdoor

本案的设计为北欧风格，北欧寒冷且湿润的海洋性气候使设计特意将室内布置成暖黄色调。

室内明显能够感受到设计师对平面图形的喜爱，尤其是方形和圆形。客厅地板用小方块铺成，沙发线条简单而造型方正，带来稳重干练的住家感受。此外，客厅顶部与餐厅的圆形水晶吊灯遥相呼应，映照着古朴深色的餐桌及包围型的座椅，与一旁的浑圆造型现代艺术品相得益彰，整个空间传递着艺术的灵感。

在宁静醇美的空间中，不论是紫红色地毯和鲜艳花朵，还是一幅橘红海报，都跳动着令人惊喜的现代元素，成为本案的一大亮点。各室的大幅落地窗不仅保持了充沛的天然照明，同时将室外自然同室内住家紧密拉近，使生机充盈在整套别墅中。

Knight Territory／爵士领地

项目地点：广州市番禺区雅居乐剑桥郡　／项目面积:约496平方米　／设计公司:力高设计（NIKKO)　／开发商：广州番禺雅居乐房地产开发有限公司　／采编：鲁晓辰

Type B villa in this case has nearly 500 square meters area. The designer uses modern costly style as the dominant element, trying to create a dream-like ideal and high-level shelter for the principal people.

Golden TV setting wall, nostalgic crystal droplight and glittering translucent glass decorations, all of which composed the splendid opening view. The black brown as the whole hue, contracted-lined but not concise sofa keeps the balance of luxury in the living room, making the space flowing elegant atmosphere and escaping from confused magic.

Sitting room and study room equipped glass as a barrier, not only increases the space outspread feeling, and the wooden bookcase gives fully sedate and the temperament of literature and art. The style of bedroom is the same as the sitting room. It is simple and clean indoor and the quiet and elegant atmosphere makes residents feel more sweet.

本案B型别墅拥有近500平方米的使用面积，设计师以现代奢华风格为主导，力求给上流人士创建一个梦幻般理想而高端的住所。

金色的电视背景墙、复古水晶吊灯以及各色晶莹剔透的玻璃配饰，如此般恢弘的开场映入眼帘。室内采用黑棕色作为整体色调，线条简约却绝不简单的沙发在客厅中拿捏着华丽的尺度，使空间流淌着华贵的气息，逃脱迷乱的魔障。

客厅与书房以玻璃作为屏障，不仅增加了空间的延伸感，且木质书橱周正的结构透出稳重而文艺的气质。卧室延续了客厅的风格，室内布置简洁，静谧优雅的氛围令住户倍感温馨。

Adoration／尊崇姿态

项目地点：广州南湖颐和高尔夫庄园　／项目面积：600-850平方米　／开发商：广州颐和地产　／采编：Y·J Lee　鲁晓辰

The design style of the neoclassicism inherits the historical and cultural essence of the classicism, and it slams the over complicated line and skin texture, meanwhile combining the modern material to present a classic and contracted new trend.

The designer transforms top villa of nearly 1000 square meters into a group's senior club, which has six layers. Each layer has different function to mix the owner's private space and the sharing time with guests together. It shows high-level life taste while getting rid of the mutual interference.

The neoclassical style of this villa reflects a pluralistic way of thinking of the designer, it combines the romantic feelings with the modern people's demand towards life quality. Indoor it mainly adopts warm colors of cream-colored, gold and brown, lines are elegant and magnificent. A wide-area use of mirror material expresses European palace legacy in this room, luxury as well as line feeling. Large area use of the black color sharply shapes the whole room into aristocratic temperament. In addition, both the lounge and the living room have commodious space, making the host and guests comfortable and enjoying in the infinite elegant atmosphere.

新古典主义的设计风格秉承着古典主义风格的历史底蕴与文化底蕴，但摒弃了过于复杂的线条和肌理，并与现代的材质相结合，呈现出古典而简约的新风尚。

设计师将这座近1000平方米的顶级别墅，设计为一家集团公司的高级俱乐部，共六层。每层功能分区各不相同，意将主人的私密空间及与宾客的共享时光相融合，体现高端生活品味的同时二者并不相互干扰。

本座别墅的新古典主义风格体现出设计师一种多元化的思考方式，将怀古的浪漫情怀与现代人对生活品质的需求相结合。室内主要以米黄、金、咖啡色等暖色为基调，线条周正而大气，大范围采用的镜面材料似乎将欧洲宫廷遗风复制于此，房间奢华却不失线条感，大面积黑色锐化了整个房间的贵族气质。此外，会客厅与休息室拥有宽阔的空间，使主人及宾客都自在惬意，尽情沉醉在这无际的优雅气氛中。

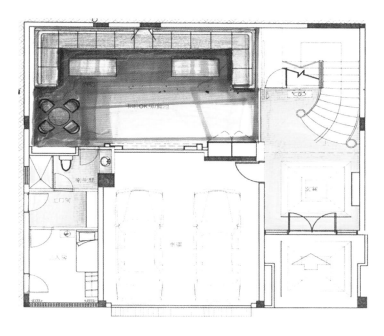

一层平面布置图

二层平面布置图

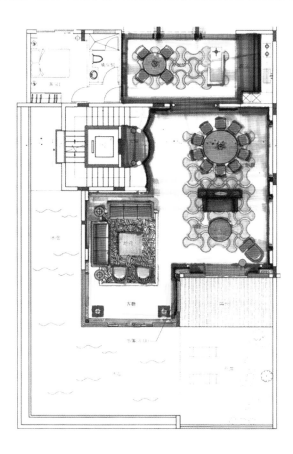

三层平面布置图

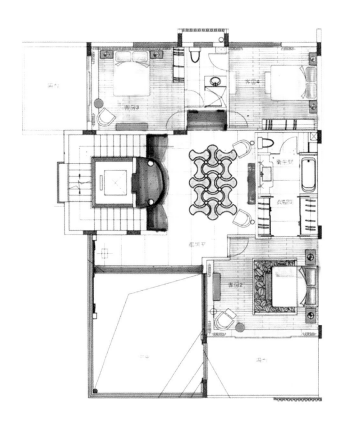

四层平面布置图

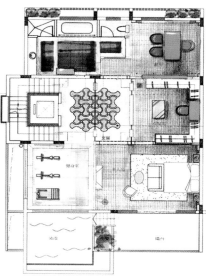

五层平面布置图

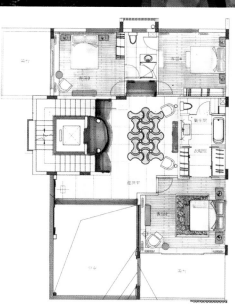

六层平面布置图

France Palace / 法兰西宫廷

项目地点：北京市朝阳区　/项目面积：2000平方米　/室内设计：广州戴维建筑设计工程有限公司　/开发商：富力地产集团　/采编：Y·J Lee

Beijing R&F Residence Villa, Type E3 is the first one introduces "France Palace" space ideology into interior design and the romantic French have never stop the pace of the pursuit of beauty, they not only require noble temperament but also artistic feelings.

The designer uses rigorous French style to interpret the logos which includes classical symmetry, archaic column, arch-shaped ceiling to show the luxurious and impressive royal atmosphere. With the elegant chandelier, lights seem to touch chords and sound a deeper note. In the dinning room, the carpet is stretching. It seems that the customs are having a forthcoming French dinner heartedly. The indoor uses large number of adorns with soft texture to set off the French style, such as excellent hard palace sofa and silk printed classical flowers, a variety of exquisite furnishings show the permeate traces.

富力公馆E3型别墅在设计中率先导入"法兰西宫廷"的空间理念,对于住宅,浪漫的法国人从未停止追求美的脚步,他们不仅要求具备贵族气质,充满艺术感的情怀也必不可少。
设计师运用严谨的法式风格来表述理念,法式元素中用经典对称、古典装饰柱、拱形天花板营造出富丽堂皇、气派庄重的宫廷气氛。造型优雅的吊灯下,光线仿佛波动着帷幔,静静弹奏着乐曲。餐厅处,拼花地毯舒展开来,令人尽情享受着一场优雅的法式晚宴。同时,室内采用了较多软饰烘托法国风情,例如质地精良的硬包宫廷沙发和古典花卉图案的真丝布料,各类精致的陈设品无不渗透出精心设计的痕迹。

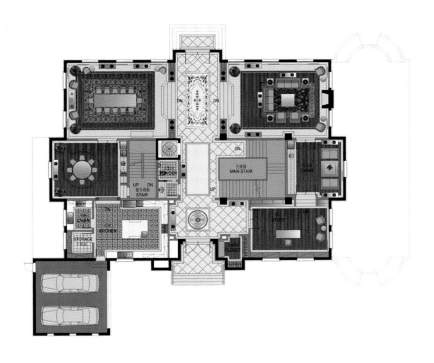

一层平面布置图

二层平面布置图

三层平面布置图

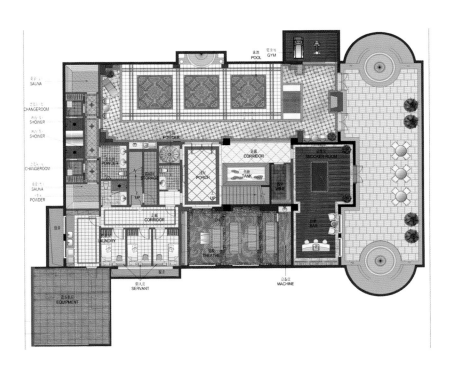

四层平面布置图

Precious Collection/臻品典藏

项目地点：广东东莞厚街　/项目面积：532平方米　/设计公司：戴勇室内设计师事务所　/开发商：丰泰集团　/采编：Y·J Lee 鲁晓辰

In this case, the designer focuses on creating a way of life with Tuscan-style, which is not only a leisurely rhythm, but also a wild and colorful space.

The case is a private mansion relying on the natural landscape, the French windows attract the blue sky and white clouds into the room, without partition of indoor and outdoor. The dinning room is just like a forest because of combination of the arc and the log ceiling. white mosaics occurring within ceiling and the LED lamp make people feels new, the different kinds of vases and ornaments exude the warmth and romance of Tuscany.

On the second floor there are two bedrooms for the guests which one is warm color to interpret tender feeling, the other is light color to interpret pure feeling. The host bedroom is on the third floor, the soft cotton, the valuable silk and the noble fur, all things are precious and beautiful. The designer boils the Tuscany style here and demonstrates the perception about art and life in this Italy town. The grains of wooden furniture stretch freely and the flowers of fabric forget blooming, all these are filled with textured touching and a collision of mind.

本案中，设计师着力于创建一种托斯卡纳式的生活方式，它既是远离现实的悠闲节奏，同时也充满着山罂粟花的狂放、艳丽。

本案是依托自然景观的私家大宅，落地窗吸引户外的蓝天白云穿堂而入，模糊了室内外的界限，餐厅因拱门和原木吊顶结为一体，宛如树影憧憧的自然森林。嵌入吊顶的洁白马赛克和LED灯管令人感觉耳目一新，而各式花瓶、摆件在细节处散发着托斯卡纳的温情与浪漫。

二层客房中，一间以暖色凝聚温情，一间以淡色寄寓纯真。独享三层空间的主人房内棉的软、丝的珍、毛绒的高贵，一切细致且精美。设计师将托斯卡纳风貌浓缩于此，尽情展现着对意大利小镇的艺术、生活的领悟。木制家具上的纹理自然地舒展，织物上花朵忘记了怒放，光影中隐约浮现着斑驳的痕迹，却赋予了触摸的质感和碰撞心灵的情趣。

Sensational Mix & Match／惊艳混搭

项目地点：中国台北　／项目面积：148平方米　／设计公司：高巢设计　／采编：Y·J Lee　鲁晓辰

The popular of the Mix and Match style mostly stems from the fashion industry in 2001. Now the wind of "Mix and Match" blow to interior design, mixing and blending a variety of style in the whole thing and finding out their advantages to design a fashion house in the modern city.

"Mix and Match" means not only adding all elements simply, but also differing primary from secondary. In this case, the modern style is the main design style, in distribution indoor each part transits naturally and meets the needs of the host's living to the greatest extent. In the sitting room the furniture arranges simply and clearly, the black-white mosaic is embellished at one corner, all kinds of materials make us to find the scenery pleasing to both the eye and the mind. The dinning room is elegant with pure white desks and chairs,matching black Baroque style droplight, the combination of black and white forms a unique mysterious feelings.

混搭的流行，更多的是源自2001年的服装界。如今，"混搭风"吹向室内设计中，将各类风格融会贯通，集各类所长打造现代都市的时尚居室。
混搭并不意味着一味的叠加，而要求主次分明、错落有致。本案以现代风格为主，室内布局中各分区过渡自然，最大程度地满足主人的居住要求。客厅处家具布置简洁明朗，黑白马赛克饰品点缀于一角，各种材质在柔和光线下令人赏心悦目。餐厅使用纯白桌椅凸显雅致，同时搭配黑色巴洛克风格吊灯，黑白相间营造出神秘气息。

Pure Beauty / 纯美乐章

项目地点：中国台北　/项目面积：200-250平方米　/设计公司：杰克文生设计　/设计师：陈茂雄　/采编：Y·J Lee 鲁晓辰

Simple and natural style is always more attractive than the artificial. The modern interior design must be conformable to client taste and meet the modern lifestyle.

The color of overall space builds clean and elegant living atmosphere, the designer uses white jazz to create a gorgeous visual enjoyment.

The indoor uses square design to devide the space, without slider between the restaurant and living room can enlarge the space visually. The large windows are beneficial to natural lighting and make the exquisite ornaments placed at each corner emit elegant and clear feeling with the transformation of light.

简约自然始终比矫揉造作更加吸引人。当今的住宅设计不仅要尊重客户喜爱的风格，同时应注重实用性，符合现代人的生活品味。

整体空间的色调铺陈出素雅的居家氛围，爵士白在设计师手中变化着表现形态，家具、饰品等同色色块相互辉映，缔造精致华丽的视觉享受。

室内采用方块设计来切割空间，餐厅与客厅之间不设隔离，尽可能放大空间的视觉效果。各室均以大窗采光，摆放于各个角落的精巧饰品随着光线的变换，散发着雅致、洁净的气息。

Sleepless City／不夜城

项目地点：上海市卢湾区马当路222号　　/项目面积：400-600平方米　　/开发商：上海安信复兴置地有限公司　　/采编：Y·J Lee　鲁晓辰

Huafu Mansion is setting in the center of Shanghai, near the elegant and advanced Huaihai Road (M) and Xin Tiandi. The Huaihai Raod(M) was called Xiafei Raod at Former French Concession, so it was under Western influence and carried on Chinese traditional style and forms the unique style combining with Chinese and Western.

This case focuses on the French esthetic philosophy and natural subject about different kinds of modalities. In the interior design, the geometric figures of representing ideal of Cubism and Assyrian temple tower originating from the ancient culture. And all kinds of molds are full of cultural intension with history precipitate. At the same time, you can find different kinds of bizarre materials like pearl, ivory, gem, enamel and ebony-free, small fig tree, the thought from designer shows the unique temperament of combination of Chinese and foreign culture and interprets the luxurious and prosperous of modern Shanghai and steady and modest deeply into the soul.

华府天地身处上海市中心，毗邻高雅繁华的淮海中路与上海新天地。淮海中路于法租界时期称"霞飞路"，受西方建筑风格影响并继承中国传统风格，形成了中西合璧的独特面貌。

本案以极度修饰的法国审美哲学以及以花饰和程式化的自然主题为特点。在室内装饰上，代表立体派理想的几何图形，来源于古代文化的亚述古庙塔，各种造型饱含着历史的沉淀后的文化内涵。同时，你可以惊喜地发现各类奇异的材料，如珍珠、象牙、宝石、珐琅及乌檀木、小无花果树等珍贵木材，设计师的考究凸显出华府天地华洋共处的独特气质，极力展现现代上海的繁华贵气与深入灵魂的沉稳谦逊。

一层平面布置图

二层平面布置图

Extreme Noble／极致尊贵

项目地点：上海市卢湾区马当路222号　　/项目面积400-600平方米　　/开发商：上海安信复兴置地有限公司　　/采编：Y·J Lee　鲁晓辰

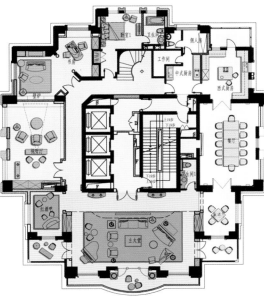

一层平面布置图

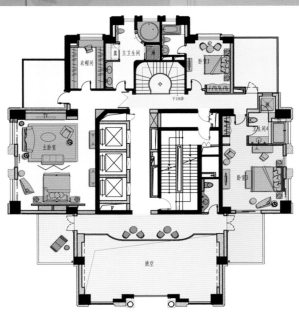

二层平面布置图

In the view of design, this case focuses on reproducing grand and imposing architecture style of the French aristocracy times and offers a polished life environment.

While panning the layout, the parlor with selected high design emphasizes the exalted manner of the villa and the arch ceiling looks like large numbers of stars surround the moon to make the crystal chandelier keep shining. Carve patterns and curves are used in detail, which on Process Technology is excellent and elegant.

The main colors are jazz white and wood in this interior design reflects the extraordinary taste of the owner. There are different designs of ceilings in each room and designer provides a cabinet for the host who enjoys tasting wine, which seems some exquisite ornaments attract people and lent strong and classical French atmosphere to this case.

本案意在重现法国贵族时代建筑的恢弘气势，为客户提供一个舒适优雅的居住环境。

在布局上，客厅以挑高设计设计凸显豪华别墅的尊贵气派，拱形顶部设计如众星拱月般使水晶吊灯拥有耀眼的璀璨光芒。细节上运用了雕花和曲线，制作工艺精细而讲究。

室内以爵士白与木色为主，优雅大方的同时体现主人不俗的艺术品味。各室均对顶部做出不同设计，细致程度可见一斑。设计师不忘为喜爱品酒的主人打造酒柜，整面墙的红酒如同精巧的饰物吸引着主人驻足把玩，更增添本案浓郁而经典的法式氛围。

Bright Starry River／星河璀璨

项目地点：上海市浦东新区锦绣路會　／项目面积:475平方米　／设计公司：邱德光设计事务所　／采编：Y·J Lee 鲁晓辰

This mansion is continued to do luxurious design style of Galaxy Bay and the designer uses some elements with individuality of Baroque style to build a modern mansion.

Stepping into the hall, there are 8 pillars before your eyes that creates a luxurious feeling against jazz white. The tea table with natural texture is majestic of starry sky, combined with black chandelier of Baroque style, the combination of black and white creates a strong and magnificent feeling.

Designers are not rigidly adhere to traditional complex style and trying to use fasion materials and geometric lines to create a modern design conception. The shining cortex is not only used on sofa but also appears on the wall, and common square patterns with raised outline gives a fashion face of elegant design by modern technology is elaborating a unique esthetic attitude.

本案为一套大户型住宅，设计风格上延续新河湾一贯的奢华风格，令人耳目一新的是设计师运用巴洛克风格的个性元素来打造现代居室。

步入大厅，八根黑云石立柱别具一格地呈现在眼前，在爵士白反衬下流露出豪门大户风范。自然纹路的茶几如星云般恢弘大气，而复古的巴洛克风格的黑色吊灯与之相互呼应，整个室内以黑白为主色调的组合给人华丽浓烈的感觉，更突显时尚利落。

设计师不拘泥于传统风格的繁复花哨，试图用时尚的材质、几何化的线条来创新现代设计理念。带有光泽质感的皮质不仅用在沙发，同时出现在卧室的墙面上，常用的方形图案有了浮凸线条，通过现代工艺赋予奢华设计时尚的面孔，阐述着一种别具风格的美学态度。

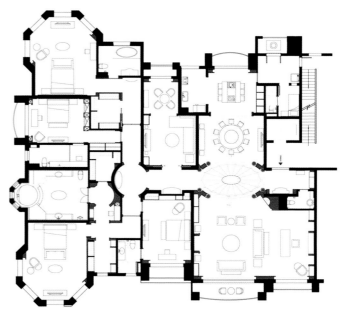

平面布置图

French Deep Feeling／法式浓情

项目地点：深圳　/项目面积：500平方米　/设计公司：PINKI（品伊）创意机构刘卫军设计师事务所　/采编：Y·J Lee　鲁晓辰

case inherits the principle of French collocation style and decorates the nature the buildings, the designer focuses on the feeling in your heart and overflows comfortable relaxation feelings.
designer devotes himself to depict the luxury of French royal and uses brown, black and dark red to interpret a noble life feeling indoor, making the villa more majesty stable. Next to the
itional French columns, deep cello is played slowly to lead people to the Champs Avenue built at 17th century by Louis XIV, when touches the classical stones listening Clip-clop, you can feel
breathing the aroma of flowers.
e parlor, the designer uses colorful landscape combined with stone and wood, not luxurious only, but adds an exotic atmosphere. The rooms are large windows, especially the French windows
oth sides in the living room are not only beneficial light, but through a hazy curtain, the scenery outside will be led to the house. In the bedroom, a touch of bright blue makes people never
et and the charming French sky is so close that you can touch it , space and time suddenly change the appearance.

秉持典型的法式风格搭配原则，将建筑点缀在自然中，设计上讲求心灵的自然回归感，给人一种迎面而来的浓郁气息。
师致力描绘出法国宫廷的恢弘大气，室内整体运用棕色、黑色、暗红等色调来表达贵族生活的精致之感，同时更具别墅的威严稳重。传统的复古法式廊柱旁，低沉的大提琴缓缓地奏响，引领人
向于17世纪建造的香榭丽舍大街——手指划过古朴的石块，久远的马蹄声划过耳畔，就这样呼吸着日光下花朵散发的香气。
内，设计师将彩色的风景绘画与石色、木色相互辉映，不再一味的奢华气派，而是使周正稳重的室内倍增异国浪漫气息。各室均为大面积窗户，尤其是客厅的两面落地窗不仅利于采光，同时透
胧的窗帘将建筑外的景致引至屋内。卧室内，一抹耀眼的蓝色让人过目不忘，法国迷人的天空如近在咫尺般可以触碰，时空转换瞬间变换了模样。

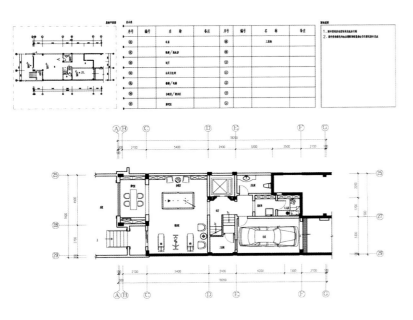

负一层平面布置图

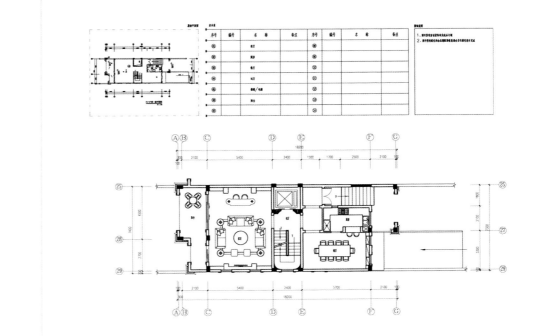

一层平面布置图

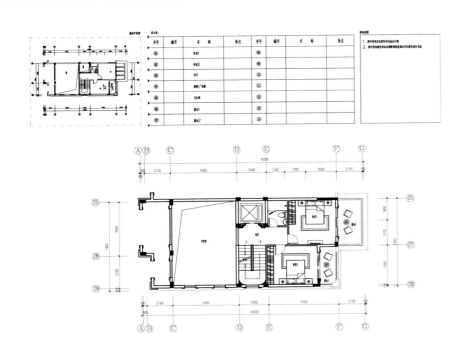

二层平面布置图

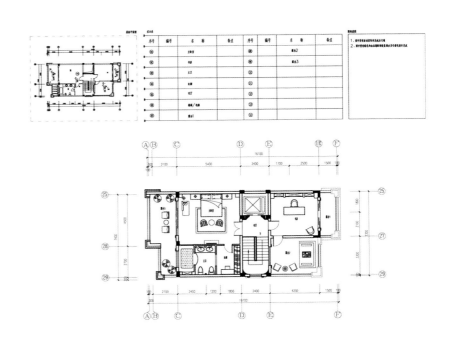

三层平面布置图

Sweet Perfume/馥郁芬芳

项目地点：西安市　　/项目面积：570平方米　　/设计公司：PINKI（品伊）创意机构刘卫军设计师事务所　　/开发商：西安深鸿基房地产开发有限公司　　/采编：鲁晓辰

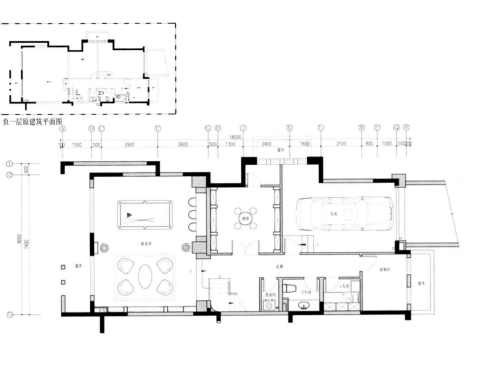

负一平面布置图

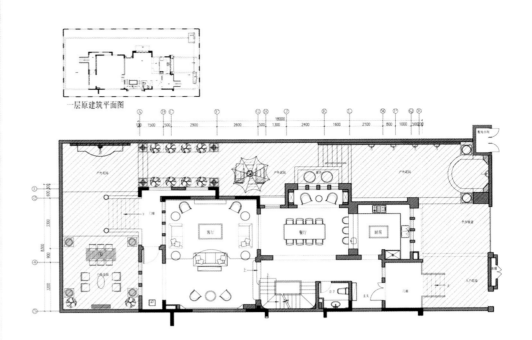

一层平面布置图

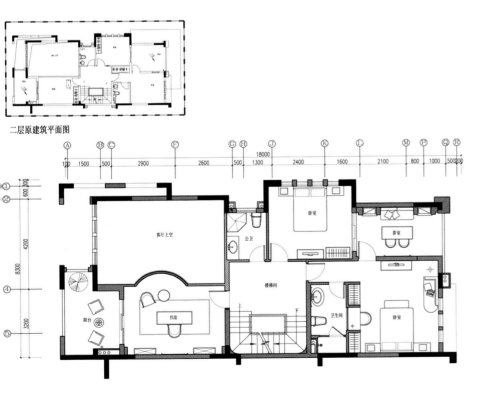

二层平面布置图

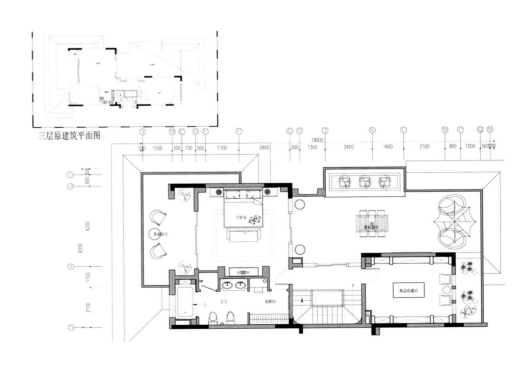

三层平面布置图

You will immediately think of the lavenders waving in the sun and some blocks of wine estate at the mention of provence, while walking here you will feel intoxicated. The theme of this case is "sweet perfume" represents the feelings of Provence, the designer tries to take people away from the busy and noisy city and lives in the distant charming European town.

This case borrows the local characteristics of Provence, in the sitting room the whole background wall is made of stones to introduce the classical feelings, the flower-pattern dark color carpet and the lamp decorated with mythological figures are tastefully appointed artworks. The design doesn't only chase for the luxury on the surface but shows the combination attitude of leisure and luxury to describe the fantastic taste of owner. The function of the villa is not only dwelling but also elevating to art form,

一提到普罗旺斯，人们必定会联想到阳光下大片随风摇曳的紫色薰衣草，一座座果实累累的葡萄酒庄园，信步其中一切都那么令人陶醉。本案主题"馥郁芬芳"，代表着浓郁的普罗旺斯的气息，设计师致力于将忙碌在喧嚣都市的人们抽离至那遥远却迷人的欧洲小镇。

本案借用普罗旺斯的当地特色，客厅处整面石砖砌成的背景墙，将古朴的美感引入屋内，花朵图案的深色沙发和神话人物装饰灯具简约中带有品味。整个设计中并未单单追求表面的纸醉金迷，而是表现一种休闲中带有奢华，兼具文化性与艺术性的非凡品位，为客户打造崇尚深层次的生活理念。别墅的功能也不仅仅限于居住，而是升华至艺术的高度。

Elegant Modern Luxury／优雅的现代奢华

项目地点：广州南湖颐和高尔夫庄园　　／项目面积：400-600平方米　　／开发商：广州颐和地产开发有限公司　　／采编：Y·J Lee　鲁晓辰

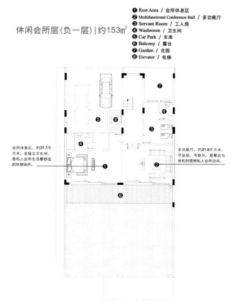

休闲会所层（负一层）｜约153㎡

❶ Rest Area ／ 会所休息区
❷ Multifunctional Conference Hall ／ 多功能厅
❸ Servant Room ／ 工人房
❹ Washroom ／ 卫生间
❺ Car Park ／ 车库
❻ Balcony ／ 露台
❼ Garden ／ 花园
❽ Elevator ／ 电梯

负一层平面布置图

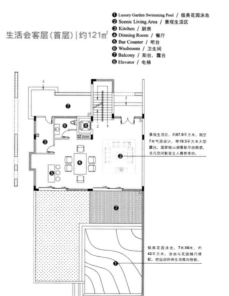

生活会客层（首层）｜约121㎡

❶ Luxury Garden Swimming Pool ／ 极美花园泳池
❷ Scenic Living Area ／ 景观生活区
❸ Kitchen ／ 厨房
❹ Dinning Room ／ 餐厅
❺ Bar Counter ／ 吧台
❻ Washroom ／ 卫生间
❼ Balcony ／ 阳台、露台
❽ Elevator ／ 电梯

一层平面布置图

家族主人层（二层）｜约104㎡

❶ Leisure Bedroom ／ 景观主卧空间
❷ Luxury Bedroom ／ 景观次卧空间
❸ Washroom ／ 卫生间
❹ Balcony ／ 露台
❺ Elevator ／ 电梯

二层平面布置图

Guangzhou Yihe Golf Manor Zoner is at the foot of the Phoenix Hill, the hill scenery is extremely beautiful. the unique hillside terrace layout with the change in mountain to make the clients feel about the natural living experience.

This case is Manhattan Business Hotel Club, the design with large spaces, mary rooms and halls can meet not only the family's comfortable life but also the casual gatherings of corporate executives. The careful design of large dinning room in second-bottom floor, which shows the clever blend of public space and private room and the feature fitted business and live is clearly.

The interior design uses the wood and warm color as main tone, combined with some furnishings with beautiful outlines to make the whole space look Fashionable and openhanded.

广州颐和高尔夫庄园伫立在凤凰山边，山景风情超凡卓群，独创的傍山露台随着山势变化错落布局，务必使客户感受到纯净无瑕的居住体验。

本案为曼哈顿商务酒店会所，多空间、多居室、多大厅的设计既可满足全家人的舒适生活，又适合企业高层的休闲聚会。负二层大型餐厅的细心设计，体现出本案公共空间和私人领域的巧妙结合，宜商宜居的功能特点体现无遗。

室内以沉稳的木色与素雅的暖色为主色调，点缀以线条华美的家具与饰品，各种风格有主有次的结合，整个空间显得大气、时尚。

Mansion Of Refined Scholar / 雅士豪宅

项目地点：广东中山 /项目面积：460平方米 /设计公司：广州市韦格斯杨设计有限公司 /采编：Y·J Lee 鲁晓辰

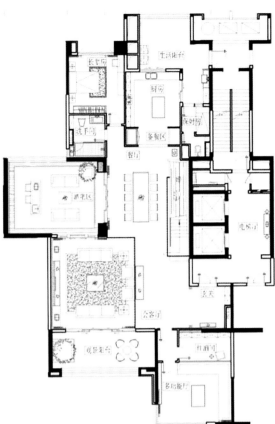

一层平面设计图

二层平面设计图

The style of this case is modern design, taking luxury as the theme. The wall and floor decorated with wood stone imported from Italy. The whole shows harmonious unity and enjoys luxurious feeling with low-key.

The compound structure extends the space for the parlor, the French windows attract the lights into the room to the greatest extent, not only make the customers feel that the space is wide and bright but also makes the TV background wall stone veins are full of texture in the sun. The bric-a-brac with trunk shape in the dinning room adds artistic atmosphere and furniture is upright which grains is similar with the TV backdrop.

The polka dots are used in each room that reflects the mix of entirety and personality, interprets the implicit aesthetic feeling like the broad starlit sky.

本案采用现代设计，以奢华大气为主题。全屋采用进口意大利的木纹石铺成墙身与地面，整体显示出和谐的统一感，低调中尽情绽放着奢华的感受。
复式结构为客厅延伸了空间，大落地窗户将光线照明最大程度地引进室内，不但能使客户感受到空间的开阔与明亮，同时使电视背景墙的石材纹理在阳光下充满质感。餐厅中树干造型的摆设增添了室内的艺术感，桌椅均采用方正造型，纹路与电视背景墙处不谋而合。
全屋各室内穿插着钢镜圆钉图案，使整体性与个性化得到合理的融合，展现出本案如星空般浩瀚而静谧的含蓄美感。

The Beauty Of Floweriness / 绚烂之美

项目地点：广州增城市新塘镇永和菱元村 /项目面积：330-500平方米 /设计公司:广州市韦格斯杨设计有限公司 /开发商：中颐集团 /采编：Y·J Lee 鲁晓辰

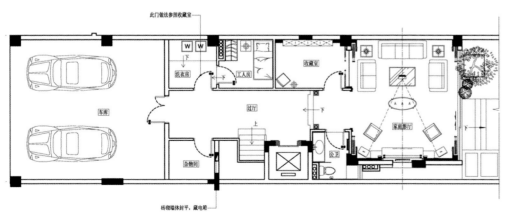

负二层平面设计图

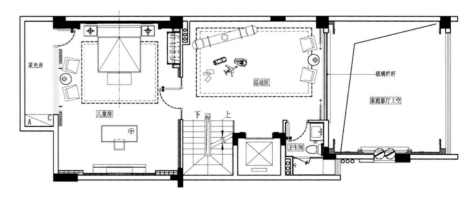

负一层平面设计图

一层平面设计图

二层平面设计图

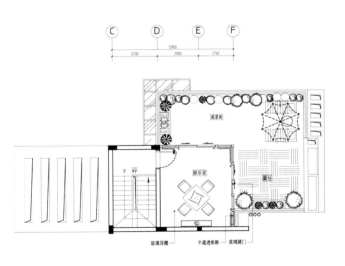

顶层平面设计图

The ideas of pursuing natural and environmental protection promote vigorously the Southeast Asian style, people are interested in natural life and the pluralism of Southeast Asian makes it different from other place.

This case is a five- floor villa and the indoor reverberates with island ecology, the light wallpaper is combined with wooden decorations and when you see beautiful colors purple and red you feel as if you can hear elephants crying and worms sing, under the broadleaf tree a charming girl's tiny mole in the forehead is sending forth the mysterious breath .

In furniture and decoration the Southeast Asian style focuses on the natural materials such as rattan-woven seats and stone accessories demonstrate the unique characteristic of tropical island.The wooden screens not just part the rooms and show the beauty of the whole. The tropical colors make you have a free feeling everyday as if you are on the vacation.

崇尚自然、追求环保的理念大力推动着东南亚风格的推广，人们日益对纯天然的生活心生向往，东南亚地区多元化的特点使之别具风情。

本案为一个五层别墅，室内洋溢着热带的东南亚岛屿特色，浅色的墙纸与木质装饰面相结合，而当目光触及紫色、红色等鲜艳的地域色彩时，整个别墅仿佛可以感觉到象鸣虫嘶的自然意境，还有阔叶树下妙龄女郎的额间红砂痣散发着的神秘气息。

家具和配饰上，东南亚风格强调使用自然的材质，如藤条座椅与石材饰物表达着热带岛屿的独有特色。木质屏风不仅实现了各室的隔断，同时体现出整体的和谐之美。生活在此，热带绚烂而浓郁的色彩令人每天都拥有如度假般的自在感受。

Art Gallery／艺廊

项目地点：北京市朝阳区　／项目面积：330平方米　／设计公司：邱德光设计事务所　／开发商：御嘉置地有限公司　／采编：Y·J Lee　鲁晓辰

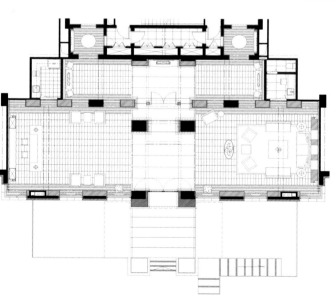

平面设计图

Beijing NAGA Mansion is explained as "art gallery", combined with Arc Deco style to make the quality of living turn to a kind of pure art taste.

The living area wide and alleys narrow forms a clear layout in the public area, which transits from the opening to the privacy. The interior design uses modern materials to create modern feeling. Ebony and biotite unprocessed natural lines are used to compose unique patterns, that fits the purpose of the owner who chases the larruping identity. The modern black is the main color, combined with the smooth color stone to interpret the artwork and create a strong cultural atmosphere, such as the sculpture by Bolin Yang.

This case has a fashion attitude, which shows the mix of classic and modern, the exquisite of Art Deco style shows the mansion's manner and we can find the Chinese aesthetic essence about the space.

设计师以"艺廊"的概念诠释北京NAGA上院大堂,并融入了Art Deco艺术风格,使居住品质蜕变为一种纯粹的居住艺术品位。
公共区域以住面宽、胡同窄形成了明确的布局,表现由开放到私密的过渡作用,并在空间中创造对称的结构。室内大量利用现代材料打造现代感。同时运用黑檀木、黑云石未经打磨的自然曲线组合出独特的图形,更符合豪宅主人追求独特性的诉求。室内整体以时尚的黑色搭配石材的温润色感,凸显空间中各项艺术陈设品,如大厅中杨柏林大师的雕塑,塑造出浓厚的文化艺术气息。
本案在整体设计上具有高贵优雅的时尚姿态,沉稳中带有新颖,Art Deco艺术风格的细腻显现出大宅的气派格局,更找到关于空间中蕴含的中式审美脉络。

Lustrous City／绚丽之都

项目地点：北京市朝阳北路　／项目面积：约300平方米　／设计公司：邱德光设计事务所　／开发商：宏宇集团　／采编：Y·J Lee　鲁晓辰

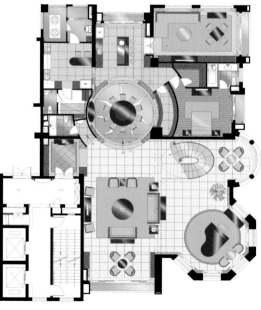

一层平面设计图

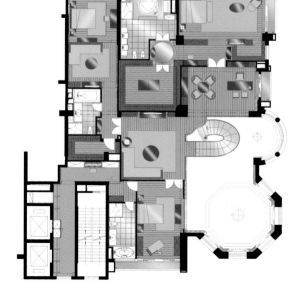

二层平面设计图

This case takes black, white and grey as the hues to create a modern luxurious mansion, the designer defines a high-quality lifestyle in his unique own style and express the pursuit of taste.

The designer uses simple colors to create an enchanted jumbo space, which shakes off the fetters of the traditional way that using gold expresses luxury, the three colors not only show the noble qualities, but also display s active lifestyle of owner's. The designer expresses luxury by the low-pitched and unique way, interpreting rationally and smartly the high-quality and romantic feeling of modern social elite group.

本案以黑、白、灰为主色打造现代奢华大宅，设计师以独特的个人风格诠释高品质的生活方式，表达着对格调与品位的追求。

设计师运用素雅的色彩营造出美轮美奂的超大空间，跳出了一贯用金色诠释奢华的传统套路，黑、白、灰三色不仅表达着淡然的高贵气质，更突显户主积极的生活态度。设计师以低调和相对专属的方式诠释奢华，以理性而睿智的态度演绎现代精英群体讲究高品质且典雅悠适的浪漫情怀。

Continental Luxury／欧陆奢华

项目地点：广州市花都　/项目面积：约450平方米　/设计公司：梁志天设计师有限公司　/设计师：梁志天　/采编：Y·J Lee　鲁晓辰

The Hong Kong designer Steven Leung has been famous for the style of "simple," working for presenting perfect space utilization, materials selection, color mixing, appropriate proportion and light arrangement. In his view, the design should reflect the character, preferences and cultural background of the owner, and shouldn't chase the surface trend and forget the deep meaning.

This case is his usual style, the designer extensive use of jazz white to create a refined atmosphere. But he changes the use of materials and shapes, such as using soft-wrappers in ceiling and large areas of patterns on the pure white wall, although they have clear distinction between uniform colors. In the space of European style Intersperse with beautiful and charming Baroque style, like black charming eyes enrich the visual impression for the whole space.

香港室内设计师梁志天的设计一直以"简约"风格著称,致力于呈现完美的空间运用、材料选择、颜色搭配、适当的比例和光线配合。在他看来,设计应该体现居住者的性格、喜好和文化背景,不应追逐表面的潮流而忘记深层次的内涵。

本案是他风格的一贯延续,设计中大量使用爵士白,营造出精致细腻的氛围。但设计师在使用中变换着材质和造型,例如将软包用至吊顶以及大面积图案铺成的纯白墙壁,虽颜色统一却区分明显。在整个充满欧陆风情的空间中穿插着华丽而迷人的巴洛克风格,它像黑色魅惑的眼睛为屋内丰富了视觉感受。

Shanghai Dragon Lake-Yanlan Hill Villa／上海龙湖滟澜山别墅

项目地点：上海青浦区　／项目面积：400-500平方米　／开发商：上海恒睿房地产有限公司　／采编：Y·J Lee　鲁晓辰

Shanghai Dragon Lake-Yanlan Hill example flat is American pastoral style which preaches the ideology --"returning to nature" and makes every effort to show a leisure and comfortable life style of countryside in the house.

In the sitting room wallpapers and wood are used largely, the furniture style is antique ,the main color of earth added with small green sofa is fresh and natural, the use of cotton and cloth combines the soft family atmosphere, and plants and flowers are put in the house not only increase the provides colorful hue but also adds a countryside scent.

The wall of the small reception room is paved with stones, the sunshine passes into the room through the windows , by the shinning skin sofa some red wine is breathing fragrantly, and what is more wonderful thing that be shared with friends in the world now?

上海龙湖滟澜山A2样板房采用美式田园风格，它倡导"回归自然"的理念，力求在室内环境中表现出休闲、舒畅的田园生活。

本案客厅运用大量的墙纸与木材，家具风格仿古，大地色的主色调搭配了绿色的小沙发显得清新自然，棉、布等材质的使用搭配出舒适柔性的家庭氛围，而巧于设置室内花朵与植物不但提供了缤纷的色调，同时更增添一份乡村气息。

小型会客室中墙面采用石材平铺，阳光透过窗户撒入室内，光亮的皮质沙发旁红酒正吐露着芬芳，这世界上又有什么比与朋友分享此时更美好的事呢？

Life Style／情调

项目地点:青浦区嘉松中路6788弄　／项目面积：约220平方米　／开发商:上海恒睿房地产有限公司　／采编：Y·J Lee 鲁晓辰

The case uses gold as the main color to create distinguished luxury atmosphere, and adds a calm and elegant feeling with black. The designer pays attention to combination of figures, for example, the decoration on the wall of living room, symmetrical ornaments and carpets are repeatedly used circular patterns; he focuses on the continuity on the materials, such as gold and silver foil that not only on the decoration frame, but also on the tea sets and ornaments, all these emphasizes its paying attention to the integrity of space.

This case is the perfect combination of Neo-classical style and European simple style, and uses many decorations of South East Asian style, which makes it elegant and practical, but reveals charming scenery of foreign country.

整个空间以金色作为主色调，奠定豪华尊贵的基调，且将黑色搭配其中以此添加一份沉稳和典雅。设计师讲究图形的呼应，如客厅内墙壁上的装饰、对称的摆设以及地毯都多次运用圆形图案，同时注重材质的照应，像金、银箔不但在镜框装饰上可见，同时用于茶具和摆件，这些都强调其注重整个空间的浑然一体。

本案可谓是新古典主义风格和欧洲简约风格的融合，并采用许多具有东南亚特色的饰品，这使奢华中兼具实用，却又不经意流露出迷人的异域风情。

Fragrance Decent / 流香暗袭

项目地点：福州　/项目面积：约150平方米　/设计公司：福州品川设计顾问有限公司　/开发商：福建泰祥房地产开发有限公司　/采编：Y·J Lee　鲁晓辰

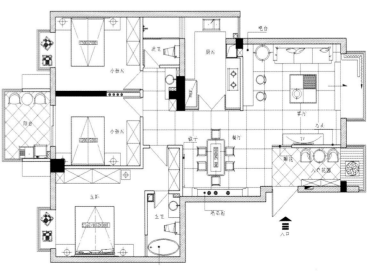

平面设计图

In the case, the designer uses modern and simple model setting against the luxurious colors such as purple and brown .Subtle and quiet mixes with free spirit, and interprets the high-quality attitude of modern life.

The interior design uses white as the main color, the hollow flowers extend the line sense for space and the large white of ceiling, kitchen and bedroom filters the vanity and affectation and adds an elegant temperament .The gorgeous shaped chandelier, complex pattern wallpaper and warm color Fu words curtain are in sunlight and lamplight, and all kinds lines and texture overlap, forming the different style, In contrast of harmonious color, which is full of humanity and artistry and provides very snug and cosy life for the residents.

本案中，设计师采用现代简约造型，佐以紫色、棕色等华贵色彩映衬，在含蓄沉静的基调下融合自由奔放，使本案尽显现代生活的优质态度。

室内主要运用白色为主色调，镂空花形隔断为空间带来线条感，吊顶、厨房、卧室的大片白色过滤了浮华和做作，更添一份高雅气质。造型华丽的吊灯、花纹繁复的壁纸以及暖色窗帘在自然光与灯光照射下，纹理交叠，生成别样风情，充满人文性与艺术感的韵律，为居住者提供着惬意与舒适。

Mirroring The Gaudiness / 镜像华美

项目地点：湖南长沙　　/项目面积：约210平方米　/设计师：朱武　/采编：Y·J Lee 鲁晓辰

This case is typically Art Deco style. Rich colors show the luxurious style and interpret a space that not only modern but also royal Art Deco style in Europe.

In order to interpret the royal complex, interior design of this case is magnificent in the extreme. Parlor and bedroom use mirror design, florid decoration lines and velvet sofa are reflected in the interlacing diamond, people are confused by where they are lived, a visionary world or real world. The designer uses straight lines many times to strengthen fluent motion in the space, which changes the material, texture and thickness, therefore we don't feel monotonously. The dome-shaped cellar, in which the design of arc display window make these bottles of wine looks like fantastic jewelry, bring to the space mix forcedly the visual concussion by using strong gold.

In the example room, the design of big windows looks out of ordinary, classic archaize bricks make the whole design don't show positively, and extend our vision.

本案为装饰艺术风格的典型代表，室内的浓郁色彩尽显豪宅风范，演绎出一个既摩登又充满欧洲皇家气息的Art Deco风格空间。

为了表达贵族情结，本案的室内设计可谓富丽堂皇。客厅及卧室吊顶采用镜面设计，交错的菱形图案倒映着繁复华丽的桌椅线条及绒面沙发，让人分不清何为虚幻，何为现实。同一空间里多次运用直线线条增加流畅感，但变换着材料、质地和线条的粗细，因此统一却不显单调。拱形圆顶的酒窖中，弧形设计的陈列柜让美酒仿佛成为了一件件炫目夺目的珍宝，金色的大量使用为空间带来强烈的视觉冲击。

整套样板间中，卧室卫生间的大窗设计别出心裁，二者风格统一，古朴仿古砖使本设计不过于浮华，并在视觉上起到绝佳的延伸作用。

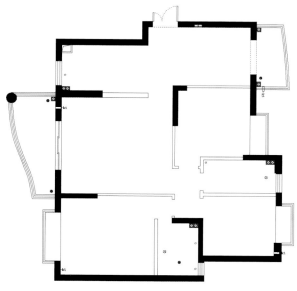

原始平面图

平面布置图

Archaic Rhymes And New Aristocrat / 古韵新贵

项目地点：湖南长沙大道508号 /项目面积：240平方米 /设计师：朱武 /开发商：湖南绿城投资置业有限公司 /采编：Y·J Lee 鲁晓辰

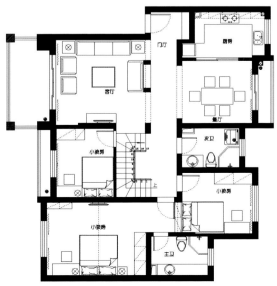

一层平面设计图

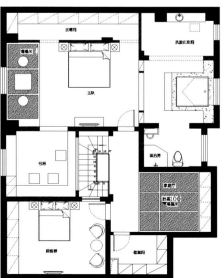

二层平面设计图

On the design theme of this case, the designer uses neo-classical style to create an ideal house by distilling and simplifying classical elements.

The sitting room is foursquare and symmetric, the furniture is put in order, the transparent stairs leads to the second floor are simple and in good taste, its glass quality is not any cramped . Stepping into the second floor, the designer uses wood and stone to build rich and classic atmosphere, classical wallpaper and warm color lights shaded into each other showing the owner's unique status and outstanding aesthetic.

在设计主题上，本案设计师采用新古典派的表现手法，通过提炼或简化古典元素来塑造理想之居。

客厅设计方正且对称，家具摆放十分规整、严谨，通向二楼的透明扶梯简洁大方，玻璃的质感在视觉上丝毫不显局促。步入二楼，设计师运用木头和石材的结合，营造厚重典雅的空间氛围，古朴的墙纸和暖色灯具更好地与木色融合，显示出业主的非凡身份与高雅审美。

The Beautiful Of Neutral Classical / 中性古典之美

项目地点：湖南长沙　/项目面积：约400平方米　/设计师：朱武　/采编：Y·J Lee

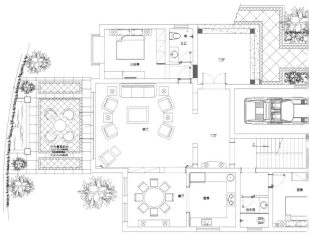

一层平面图

二层平面图

三层平面图

Classical beauty is the focal point of this space, to present deferent kind of space circumstance by the matrial change. The wall face is high color imitated old wall paper, the top is decorated with variable wood, the light used the copper-colored chandelier, the grandur feature of the space such appeared and more silent to be fit with the subject's requirement by the processing of high colour, the shape of the primitive staircase made the disign of the space to be more peaceful, noble and cultural. Bronze-colored iron railing not only distinct the separated space but also effect well as a decoration. The whole room is filled with noble spirit plused the hanging drop of the wall face and the right adornment of furniture.

古典美是这个空间的重点,通过材质的变化,营造出不同的空间氛围。墙面搭配深色仿旧的墙纸,顶面加以木材的变化,灯具用铜色吊灯,使空间的高贵气质得以体现,而深色调的处理,使空间更加安静,符合主题的需要,古朴楼梯的造型,使空间的设计更显沉稳,高贵而儒雅,多处使用的古铜色铁艺栏杆不仅区隔了空间分区,而且也起到了很好装饰效果,地面仿古砖的铺贴也很讲究,无论是在色彩还是质地上都透出一种与生俱来的高贵。再加上墙面的挂饰以及家具恰到好处的装点,使整个空间尽显高贵气质。

The Monumental Beauty/传世之美

项目地点：湖南湘潭　/项目面积：1080平方米　/设计师：谭金良　/采编：Y·J Lee

This villa has 3 stair, it covers almost 2000 square meters. Its basic concrete style is the minimalist european, do away with the most complicated lines except the many levels and sinuous wall face. The front yard used the winehouse style parking design, with a unconspicious significant honour. Courtyard designing, waterside pavillions, flower birds and insects, there are no shaort of the fundamental element of the chinese horticultural design, and be particular about the decoration. The emotional water flowes from its streaming back. So a pool of water in front of the vestibule means wealth flows in you. The old people take care of the plants and their mind of living in seclution is plain to see.

这栋别墅共有三层，占地面积近2000平方米。以简欧的建筑风格为主，抛却欧式最为复杂的线条，只留下多层次的平面和蜿蜒的墙面。前庭采用了酒店式泊车设计，有着一种不彰自显的尊贵。与之对应的是极富中式情调的庭院设计，亭台水榭，花草虫鸟，中式园林设计里的基本元素一个都不少，但对其的铺陈却极有讲究。欲水之有情，喜其回环朝穴。在前厅的前方聚一方水，也就是聚敛了财气。老人们侍弄些花花草草，颐养天年，归隐之心，昭然若见。

The two opened heavy door witnessed a flourishing family .The 10m living room can be described as beyond the momentum.The kitchen and dinning hall is of important propor in the square of floor 1.The eating people of the master's whole family can be count more than 10,so the design of the kitchen and dinning hall is also considered about the spac daily meal beside the following of american style to separate the kitchen to middle and west one to make it conveninet with generation's dinning requirment.The advan by degree space from the vestibuel to courtyard ended in the back room implied the meaning of step by step in feng shui doctrine.Central nave hall for worshipping buddha and meeting ro respectively ,regard the reminicen to the their ancestor and their belief, and provided a more private space for the master and his client's meeting .

The villa's living room is distributed to second floor and third floor .The second is grouped by 3 flat .It delivered different kind of feeling though they are all in the framework american style.He rusticity of parent's room , the steadiness of the master's room and the easiness of the chidren's room ,change of suspended ceiling and the choice of v paper , the dipartition of space and the putie tile of washroom arosed from different keyword .Little as the desk lamp or big the overall furniture ,which are made acroding to inhabitant's character,hobby and height.From the hallway to the study room, from the living room to the washroom , dressing room , tea room storage room are all included. space is designed by uncompromised attention , which made everyone feel comfortable ,and the final decoration manifested a casulity, put the touching body into immaterial s Sterotyped audio-visual room and the noise-absorbing bud pattern are responded with each other the gymnasium to sway your sweat , the underground wine cellar to keep y own red wine , the cardroom to enjoy your leisure time , which give us a deep belief that the description of luxury is on one point means waste , on the other point , priviege.

推开两扇厚重的大门，见证了一个人丁兴旺的家族。层高近10米的客厅，可以用气势恢弘来形容。厨房餐厅占据了一楼面积很大的比重。主人一家每日就餐的人员达十人之上，厨房餐厅的设计在沿袭美式风格的基础上还考虑了日常备餐的空间，把厨房分为中厨、西厨两个区域，方便几代人不同的饮食需求。从前厅到中庭再到后堂，空间呈有层次的递印证风水学说里步步高升之意。分设中厅一侧的佛堂和会客室，既考虑了业主对先辈们的缅怀之情和宗教信仰，也为主人会见客人提供了一个较为私密的空间。

别墅的寝居空间分布在二楼和三楼。二楼由三个套房组成。虽然都是在美式风格的大框架里，可是却又做出了不同的感觉。父母房的质朴，主人房的稳健，还有小孩房的轻松，同的关键词引发出吊顶的变化、墙纸的选用，空间的划分以及卫生间瓷砖的铺贴。大到整体家具小到一盏台灯，都是根据居住者的性格、喜好、身高来定制。每个套房的功能性当齐全，从玄关到书房、从卧房到卫生间以及更衣室甚至是茶水间收纳室都一一囊括。空间的精雕细琢让人最为舒适，而最终的雕琢表现出来的其实又是一种随意，不故意，寓于无形中。超立体功能的视听室，吸音的花苞图案彼此对称。挥洒汗水的健身房，私藏红酒的地下酒窖，休闲娱乐的桥牌室，这些都让我们相信，奢华意味着空间的独享。

地下室平面图

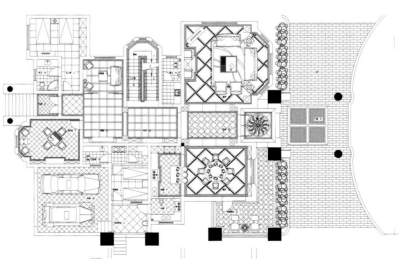

一层平面图

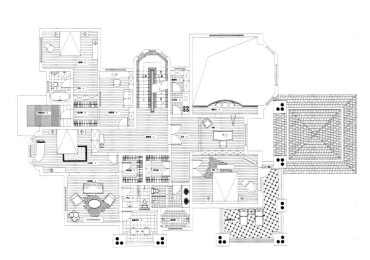

二层平面图

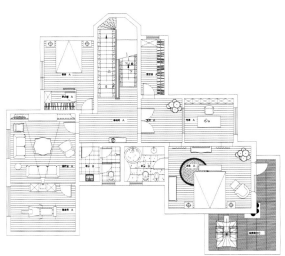

三层平面图